SOLID WASTE AS A RENEWABLE RESOURCE
Methodologies

SOLID WASTE AS A RENEWABLE RESOURCE
Methodologies

Edited by
Jimmy Alexander Faria Albanese, PhD and M. Pilar Ruiz, PhD

Apple Academic Press Inc.
3333 Mistwell Crescent
Oakville, ON L6L 0A2
Canada

Apple Academic Press Inc.
9 Spinnaker Way
Waretown, NJ 08758
USA

©2016 by Apple Academic Press, Inc.

Exclusive worldwide distribution by CRC Press, a member of Taylor & Francis Group

No claim to original U.S. Government works

Printed in the United States of America on acid-free paper

International Standard Book Number-13: 978-1-77188-243-9 (Hardcover)

This book contains information obtained from authentic and highly regarded sources. Reprinted material is quoted with permission and sources are indicated. Copyright for individual articles remains with the authors as indicated. A wide variety of references are listed. Reasonable efforts have been made to publish reliable data and information, but the authors, editors, and the publisher cannot assume responsibility for the validity of all materials or the consequences of their use. The authors, editors, and the publisher have attempted to trace the copyright holders of all material reproduced in this publication and apologize to copyright holders if permission to publish in this form has not been obtained. If any copyright material has not been acknowledged, please write and let us know so we may rectify in any future reprint.

Trademark Notice: Registered trademark of products or corporate names are used only for explanation and identification without intent to infringe.

Library and Archives Canada Cataloguing in Publication

Solid waste as a renewable resource : methodologies / edited by Jimmy Alexander Faria Albanese, PhD and M. Pilar Ruiz, PhD.

Includes bibliographical references and index.
ISBN 978-1-77188-243-9 (bound)

1. Refuse as fuel. 2. Waste products as fuel. 3. Refuse and refuse disposal. I. Faria Albanese, Jimmy Alexander, author, editor II. Ruiz, M. Pilar, author, editor

TD791.S64 2015   628.4'4   C2015-902546-X

Library of Congress Cataloging-in-Publication Data

Solid waste as a renewable resource : methodologies / editor, Jimmy Alexander Faria Albanese, PhD and M. Pilar Ruiz, PhD.

pages cm
Includes bibliographical references and index.
ISBN 978-1-77188-243-9 (alk. paper)

1. Recycling (Waste, etc.) 2. Refuse and refuse disposal--Management. 3. Compost. 4. Waste products as fuel I. Faria Albanese, Jimmy Alexander. II. Ruiz, M. Pilar (Maria Pilar)

TD794.5.S657 2015   628.4'458--dc23   2015013042

Apple Academic Press also publishes its books in a variety of electronic formats. Some content that appears in print may not be available in electronic format. For information about Apple Academic Press products, visit our website at **www.appleacademicpress.com** and the CRC Press website at **www.crc-press.com**

About the Editors

JIMMY ALEXANDER FARIA ALBANESE, PhD

Jimmy Faria is Senior Scientist at Abengoa Research, a R&D division of Abengoa. He is a chemical engineer and obtained a PhD from the University of Oklahoma (USA) in 2012. His research at the School of Chemical, Biological and Material Science at the University of Oklahoma (USA) is focused on the catalytic conversion of biomass-derived compounds in a novel nanoparticle-stabilized emulsion system developed in this group, as well as on the synthesis, characterization and applications of amphiphilic nanohybrids (e.g., enhanced oil recovery).

M. PILAR RUIZ, PhD

Maria Pilar Ruiz-Ramiro is Senior Scientist at Abengoa Research, a R&D division of Abengoa. She is a chemical engineer and obtained a PhD from the University of Zaragoza (Spain) in 2008. She later worked as Research Associate with Daniel E. Resasco at the School of Chemical, Biological and Material Science at the University of Oklahoma (USA). Her research is focused on the thermochemical conversion of biomass, synthesis and characterization of carbon solids (carbon nanotubes, biomass char and soot), and the development of nanostructured catalysts for biofuels upgrading reactions.

Contents

Acknowledgment and How to Cite ix
List of Contributors xi
Introduction xvii

Part I: Foundations

1. **Energy Recovery from Municipal and Industrial Wastes: How Much Green?** 3
   Satinder Kaur Brar, Saurabh Jyoti Sarma, and Mausam Verma

2. **Energy Recovery Potential and Life Cycle Impact Assessment of Municipal Solid Waste Management Technologies in Asian Countries Using ELP Model** 7
   Andante Hadi Pandyaswargo, Hiroshi Onoda, and Katsuya Nagata

Part II: Anaerobic Digestion

3. **Utilization of Household Food Waste for the Production of Ethanol at High Dry Material Content** 35
   Leonidas Matsakas, Dimitris Kekos, Maria Loizidou, and Paul Christakopoulos

4. **Production of Fungal Glucoamylase for Glucose Production from Food Waste** 55
   Wan Chi Lam, Daniel Pleissner, and Carol Sze Ki Lin

Part III: Composting

5. **Changes in Selected Hydrophobic Components During Composting of Municipal Solid Wastes** 73
   Jakub Bekier, Jerzy Drozd, Elżbieta Jamroz, Bogdan Jarosz, Andrzej Kocowicz, Karolina Walenczak, and Jerzy Weber

6. **Transforming Municipal Waste into a Valuable Soil Conditioner through Knowledge-Based Resource-Recovery Management** 87
   Golabi MH, Kirk Johnson, Takeshi Fujiwara, and Eri Ito

Part IV: Pyrolysis and Chemical Upgrading

7. Furfurals as Chemical Platform for Biofuels Production........ 103

Daniel E. Resasco, Surapas Sitthisa, Jimmy Faria, Teerawit Prasomsri, and M. Pilar Ruiz

Part V: Incineration and Carbonization

8. Incineration of Pre-Treated Municipal Solid Waste (MSW) for Energy Co-Generation in a Non-Densely Populated Area 147

Ettore Trulli, Vincenzo Torretta, Massimo Raboni, and Salvatore Masi

9. Gaseous Emissions During Concurrent Combustion of Biomass and Non-Recyclable Municipal Solid Waste 171

René Laryea-Goldsmith, John Oakey, and Nigel J. Simms

10. Environmental Effects of Sewage Sludge Carbonization and Other Treatment Alternatives... 189

Ning-Yi Wang, Chun-Hao Shih, Pei-Te Chiueh, and Yu-Fong Huang

Part VI: Gasification

11. An Experimental and Numerical Investigation of Fluidized Bed Gasification of Solid Waste 213

Sharmina Begum, Mohammad G. Rasul, Delwar Akbar, and David Cork

12. Gasification of Plastic Waste as Waste-to-Energy or Waste-to-Syngas Recovery Route... 241

Anke Brems, Raf Dewil, Jan Baeyens, and Rui Zhang

Author Notes.. 265

Index.. 269

Acknowledgment and How to Cite

The editor and publisher thank each of the authors who contributed to this book. The chapters in this book have been published previously. To cite the work contained in this book and to view the individual permissions, please refer to the citation at the beginning of each chapter. Each chapter was read individually and carefully selected by the editor; the result is a book that provides a nuanced look at the reuse of solid waste. The chapters included are broken into six sections, which describe the following topics:

- Chapter 1 was chosen as a good introduction to the book topic.
- Chapter 2 uses environmental load point methodology to assess best solid waste management strategies for three Asian countries, while also using a life cycle assessment to measure the environmental impact and energy recovery potential of each waste treatment scenario.
- Chapter 3 is a good analysis of best methodologies for utilizing household food wastes to achieve efficient ethanol production .
- The research done in Chapter 4 shows the great potential for using glocoamylase for food waste treatment at an industrial level while creating sustainable chemicals production.
- Chapter 5 is a solid investigation of how the biotransformation of solid wastes during composting processes may be influenced by other factors such as temperature, oxygenation, and mineral composition.
- Chapter 6 Uses the solid waste management system on the island of Guam as a microcosmic perspective with that of Isfahan, Iran, as a more macrocosmic view to reach important implications regarding the role of societal input in creating an efficient composting scenario.
- In Chapter 7 we compare different conversion strategies involving furfural and hydroxymethyl furfural, two chemical building blocks for the production of transportation fuels as well as a variety of useful acids, aldehydes, alcohols, and amines. We concluded that by varying the catalyst composition, as well as reaction conditions, a rich variety of products with different fuel properties can be obtained.

- Chapter 8 provides a solid methodological approach to determine technical solutions for managing particular municipal solid wastes (using current and future trends for composition and quantity) to define the optimal mechanical pretreatments to achieve the best-case balance between greenhouse gas emissions and heat-and-power production.
- Chapter 9 is a sound analysis of the rejected material from recycling facilities that suggests that energy conversion of such waste is practical, and that differences in geography and seasonal changes can be tolerated as they are integrated with varying fuel types.
- Chapter 10 applies life cycle assessment to a simulated sewage sludge carbonization process to evaluate the environmental effects and benefits of the carbonization process, comparing results for direct landfills, co-incineration with municipal solid waste, and mono-incineration, with the conclusion that carbonization was the most preferable sludge-handling option overall, despite the fact that the co-incineration option emitted less greenhouse gases than carbonization due to higher energy recovery ratio.
- Chapter 11 develops a model to assist researchers and professionals identify optimized conditions for solid waste gasification.
- In Chapter 12, Brems and colleagues make the case that the gasification of plastic solid waste to produce syngas is a valid recycling route

List of Contributors

Delwar Akbar
School of Business and Law, Central Queensland University, Rockhampton, QLD 4702, Australia

Jan Baeyens
College of Life Science and Technology, The Key Laboratory of Bioprocess of Beijing, Beijing University of Chemical Technology, Beijing, China

Sharmina Begum
School of Engineering and Technology, Central Queensland University, Rockhampton, QLD 4702, Australia

Jakub Bekier
Wroclaw University of Environmental and Life Sciences, Institute of Soil Science and Environmental Protection, Grunwaldzka 53, Wroclaw, Poland

Satinder Kaur Brar
Institut National de la Recherche Scientifique, Centre - Eau Terre Environnement, 490, rue de la Couronne, Québec, Québec G1K 9A9, Canada

Anke Brems
Department of Chemical Engineering, Chemical and Biochemical Process Technology and Control Section, Katholieke Universiteit Leuven, Heverlee, Belgium

Pei-Te Chiueh
Graduate Institute of Environmental Engineering, National Taiwan University, No.71 Chou-Shan Road, Taipei 106, Taiwan

Paul Christakopoulos
Department of Civil, Biochemical and Chemical Process Engineering, Division of Sustainable Process Engineering, Environmental and Natural Resources Engineering, Luleå University of Technology, SE 971 87 Luleå, Sweden

David Cork
The Corky's Group, Mayfield, NSW 2304, Australia

Raf Dewil
Department of Chemical Engineering, Chemical and Biochemical Process Technology and Control Section, Katholieke Universiteit Leuven, Heverlee, Belgium

Jerzy Drozd
Wroclaw University of Environmental and Life Sciences, Institute of Soil Science and Environmental Protection, Grunwaldzka 53, Wroclaw, Poland

Jimmy Faria
School of Chemical, Biological, and Materials Engineering, and Center for Biomass Refining University of Oklahoma, Norman, OK 73019, USA

Takeshi Fujiwara
Solid Waste Management Research Center, Okayama University, 3-1-1 Tsushima Naka, Okayama, 7008530, Japan

Yu-Fong Huang
Graduate Institute of Environmental Engineering, National Taiwan University, No.71 Chou-Shan Road, Taipei 106, Taiwan

Eri Ito
Solid Waste Management Research Center, Okayama University, 3-1-1 Tsushima Naka, Okayama, 7008530, Japan

Elżbieta Jamroz
Wroclaw University of Environmental and Life Sciences, Institute of Soil Science and Environmental Protection, Grunwaldzka 53, Wroclaw, Poland

Bogdan Jarosz
Department of Chemistry, Wroclaw University of Environmental and Life Sciences, Norwida 25/27, Wroclaw, Poland

Kirk Johnson
College of Natural and Applied Sciences, University of Guam, Mangilao, Guam 96923, USA

Dimitris Kekos
Biotechnology Laboratory, School of Chemical Engineering, National Technical University of Athens, 5 Iroon Polytechniou Str, Zografou Campus, 15780 Athens, Greece

Andrzej Kocowicz
Wroclaw University of Environmental and Life Sciences, Institute of Soil Science and Environmental Protection, Grunwaldzka 53, Wroclaw, Poland

List of Contributors

Wan Chi Lam
School of Energy and Environment, City University of Hong Kong, Tat Chee Avenue, Kowloon, Hong Kong, China

René Laryea-Goldsmith
Energy Technology Centre, Cranfield University, Cranfield, Bedfordshire, UK

Carol Sze Ki Lin
School of Energy and Environment, City University of Hong Kong, Tat Chee Avenue, Kowloon, Hong Kong, China

Maria Loizidou
Unit of Environmental Science and Technology, School of Chemical Engineering, National Technical University of Athens, 5, Iroon Polytechniou Str, Zografou Campus, 15780 Athens, Greece

Salvatore Masi
School of Engineering, University of Basilicata, Viale dell'Ateneo Lucano 10, Potenza I-85100, Italy

Leonidas Matsakas
Biotechnology Laboratory, School of Chemical Engineering, National Technical University of Athens, 5 Iroon Polytechniou Str, Zografou Campus, 15780 Athens, Greece and Department of Civil, Biochemical and Chemical Process Engineering, Division of Sustainable Process Engineering, Environmental and Natural Resources Engineering, Luleå University of Technology, SE 971 87 Luleå, Sweden

Golabi MH
College of Natural and Applied Sciences, University of Guam, Mangilao, Guam 96923, USA

Katsuya Nagata
Graduate School of Environment and Energy Engineering, Onoda/Nagata Laboratory, Waseda University, Saitama, 367-0035, Japan

John Oakey
Energy Technology Centre, Cranfield University, Cranfield, Bedfordshire, UK

Hiroshi Onoda
Graduate School of Environment and Energy Engineering, Onoda/Nagata Laboratory, Waseda University, Saitama, 367-0035, Japan

Andante Hadi Pandyaswargo
Graduate School of Environment and Energy Engineering, Onoda/Nagata Laboratory, Waseda University, Saitama, 367-0035, Japan

Daniel Pleissner
School of Energy and Environment, City University of Hong Kong, Tat Chee Avenue, Kowloon, Hong Kong, China

Teerawit Prasomsri
School of Chemical, Biological, and Materials Engineering, and Center for Biomass Refining University of Oklahoma, Norman, OK 73019, USA

Massimo Raboni
Department of Biotechnologies and Life Sciences, Insubria University, Via G.B. Vico 46, Varese I-21100, Italy

Mohammad G. Rasul
School of Engineering and Technology, Central Queensland University, Rockhampton, QLD 4702, Australia

Daniel E. Resasco
School of Chemical, Biological, and Materials Engineering, and Center for Biomass Refining University of Oklahoma, Norman, OK 73019, USA

M. Pilar Ruiz
School of Chemical, Biological, and Materials Engineering, and Center for Biomass Refining University of Oklahoma, Norman, OK 73019, USA

Saurabh Jyoti Sarma
Institut National de la Recherche Scientifique, Centre - Eau Terre Environnement, 490, rue de la Couronne, Québec, Québec G1K 9A9, Canada

Chun-Hao Shih
Graduate Institute of Environmental Engineering, National Taiwan University, No.71 Chou-Shan Road, Taipei 106, Taiwan

Nigel J. Simms
Energy Technology Centre, Cranfield University, Cranfield, Bedfordshire, UK

Surapas Sitthisa
School of Chemical, Biological, and Materials Engineering, and Center for Biomass Refining University of Oklahoma, Norman, OK 73019, USA

Vincenzo Torretta
Department of Biotechnologies and Life Sciences, Insubria University, Via G.B. Vico 46, Varese I-21100, Italy

Ettore Trulli
School of Engineering, University of Basilicata, Viale dell'Ateneo Lucano 10, Potenza I-85100, Italy

List of Contributors

Mausam Verma
CO2 Solutions Inc., 2300, rue Jean-Perrin, Québec, Québec G2C 1T9, Canada

Karolina Walenczak
Wroclaw University of Environmental and Life Sciences, Institute of Soil Science and Environmental Protection, Grunwaldzka 53, Wroclaw, Poland

Ning-Yi Wang
Graduate Institute of Environmental Engineering, National Taiwan University, No.71 Chou-Shan Road, Taipei 106, Taiwan

Jerzy Weber
Wroclaw University of Environmental and Life Sciences, Institute of Soil Science and Environmental Protection, Grunwaldzka 53, Wroclaw, Poland

Rui Zhang
College of Life Science and Technology, The Key Laboratory of Bioprocess of Beijing, Beijing University of Chemical Technology, Beijing, China

Introduction

The current environmental crisis and the shortage of fossil fuels have directed the world's attention toward other forms of energy that are more environmentally friendly and renewable, such as bio-ethanol and other biofuels. In contrast to fossil fuels, biomass-derived fuels can be CO_2-neutral, since the CO_2 produced in their combustion can be reabsorbed by green plants and algae during photosynthesis [1–4]. First-generation biofuels proved to bring with them their own set of problems relating to food shortages and rising food prices. Second-generation biofuels, however, turned to other feedstock sources.

Technical and economical challenges remain that have delayed the application of this technology to the commercial scale. The availability of enough quantities of biomass is a key factor in scaling up biofuel production, but with the development of efficient technologies and methodologies, there appear to be answers readily at hand to meet the challenge. In the United States, it is estimated that thousands of metric tons of dry biomass/year could be sustainably produced for biofuels, without a significant impact on human food, livestock feed, or export demands [5]. This amount of biomass would represent an energy equivalent that would account for 54 percent of the current annual demand of crude oil in the United States alone [6,7]. An economically sustainable and competitive process will require versatility to accept different kinds of biomass feedstocks and the ability to produce low-oxygen, high-energy content liquids, which should be fungible with conventional fuels [8].

Municipal solid waste is a particularly attractive source of endless feedstock for biofuel production. Between 2007 and 2011, global generation of municipal waste was estimated to have risen more than 37 percent, equivalent to roughly an 8% increase per year [9], a trend that promises to continue on into the twenty-first century. Reusing this material as a fuelstock will effectively "kill two birds with one stone." Other uses for solid waste, such as fertilizers, also offer potentially attractive solutions to one of the world's current greatest challenges.

Solid waste can become a viable renewable resource through a range of waste management technologies. Each have individual applications, efficiencies, and challenges. We have included here research into anaerobic digestion, composting, pyrolysis and chemical upgrading, incineration and carbonization, and gasification.

In chapters 3 and 4, Matsakas et al. and Lam et al. investigate aspects of anaerobic digestion applicable to food waste management.

One of the most practical ways to utilize municipal solid waste is composting, thereby producing materials that may be productively used to improve soil properties; chapters 5 and 6 (Bekier et al. and Mh et al.) review composting methodologies for transforming solid waste into reusable materials.

In chapter 7, our co-authors and ourselves examined furfurals as a chemical platform within the pyrolysis process for biofuel production.

Combustion, which is a rapid chemical reaction of two or more substances (more commonly called burning), offers an efficient method of electricity and heat generation using renewable energy resources. Concurrent combustion of biomass and municipal solid waste can make effective contributions toward the ongoing international commitments to minimize environmental damage [10]. In chapters 8 through 10, Trulli et al., Laryea-Goldsmith et al., and Wang et al. look at various treatments applied to combustion and incineration for managing municipal solid wastes, plastic, and sewage sludge.

Like pyrolysis and unlike combustion, gasification heats the fuel with little or no oxygen, producing "syngas," which can be used to generate energy or as a feedstock for producing methane, chemicals, biofuels, or hydrogen. In chapter 11, Begum et al. develop an optimal model for gasification of solid waste, and in chapter 12, Brems et al. examine the gasification of plastic as an effective waste-to-energy route.

Ongoing research into these methodologies is vital for the economic and environmental future of our world. There will be no single solution to the world's demand for energy, nor will we find a comprehensive technique for mitigating climate change in the years to come. Instead, we must work in multiple directions at once. The articles included within this compendium have been selected with that goal in mind. Ultimately, of course, any good management system also must involve not only governments and industries, but individuals as well. At every level, we must become aware of the environmental benefits and of the reduced danger to health and economy that results from efficient solid waste management [11,12].

REFERENCES

1. Wyman, C.E.; Hinman, N.D. Appl. Biochem. Biotechnol. 1990, 24/25, 735–753.
2. Wyman, C.E. Appl. Biochem. Biotechnol. 1994, 45/46, 897–915.
3. Tyson, K.S. Fuel Cycle Evaluations of Biomass-Ethanol and Reformulated Gasoline; Report No. NREL/TP-263-2950, DE94000227, National Renewable Energy Laboratory: Golden, CO, 1993.
4. Lynd, L.R.; Cushman, J.H.; Nichols, R.J.; Wyman, C.E. Science 1991, 251, 1318–1323.
5. Perlack, R.D.; Wright, L.L.; Turhollow, A.; Graham, R.L.; Stokes, B.; Erbach, D.C. Biomass as Feedstock for a Bioenergy and Bioproducts Industry: The Technical Feasibility of a Billion-Ton Annual Supply, Report No. DOE/GO-102995-2135; Oak Ridge National Laboratory: Oak Ridge, TN, 2005; http://www.osti.gov/ bridge.
6. Klass, D.L. Biomass for Renewable Energy, Fuels and Chemicals; Academic Press: San Diego, 1998.
7. Energy Information Administration Annual Energy Outlook 2005; Report. No. DOE/EIA-0383; U.S. Department of Energy: Washington, DC, 2006; http://www.eia.doe.gov.
8. U.S. Department of Energy, Feedstock Composition Gallery, Washington, DC, 2005; http://www.eere.energy.gov/biomass/feedstock_glossary.html.
9. United Nations Environmental Program. Developing Integrated Solid Waste Management (ISWM) Plan Training Manual. Division of Technology, Industry and Economics, International Environmental Technology Centre Osaka, Shiga Japan, 2009, 4:1-172.
10. DEFRA: UK Biomass Strategy. In Tech rep. Department for Food, Energy and Rural Affairs, London; 2007.
11. Di Mauro, C.; Bouchon, S.; Torretta, V. Industrial risk in the Lombardy Region (Italy): What people perceive and what are the gaps to improve the risk communication and the partecipatory processes. Chem. Eng. Trans. 2012, 26, 297–302.
12. Morris, M.W.; Su, S.K. Social psychological obstacles in environmental conflict resolution. Am. Behav. Sci. 1999, 42, 1322–1349.

Jimmy Alexander Faria Albanese and M. Pilar Ruiz

Chapter 1, by Brau and colleagues, describes various methods of transforming waste to fuels, as well as some pluses and minuses of each method.

Natural resource scarcity and the effects of environmental destruction have pushed societies to use and reuse resources more efficiently. Waste should no longer be seen as a burden but rather as another source of material such as energy fuel. Chapter 2, by Pandyaswargo and colleagues, analyzes the potential of three waste management scenarios that include the combination of four waste management technologies—incineration with energy recovery, composting, anaerobic digestion, and sanitary landfill gas collection—as ways to recover energy and material from municipal solid waste. The study applies the environmental load point (ELP) method and utilizes municipal waste characteristics and composition from India, Indonesia, and China as case studies. The ELP methodology employs integrated weighting in the quantification process to get a one-unit result. This study particularly uses analytic hierarchical process questionnaires to get the weighting value of the nine impact categories: energy depletion, global warming, ozone depletion, resource consumption, ecosystem influence, water pollution, waste disposal, air pollution, and acid rain. The results show that the scenario which includes composting organic waste and sanitary landfill with gas collection for energy recovery has medium environmental impact and the highest practicability. The optimum material and energy potential is from the Chinese case study in which 254 tonnes of compost fertilizer and 60 MWh of electricity is the estimated output for every 1,000 tonnes of waste treated.

Environmental issues and shortage of fossil fuels have turned the public interest to the utilization of renewable, environmentally friendly fuels, such as ethanol. In order to minimize the competition between fuels and food production, researchers are focusing their efforts to the utilization of wastes and by-products as raw materials for the production of ethanol. household food wastes are being produced in great quantities in European Union and their handling can be a challenge. Moreover, their disposal can cause severe environmental issues (for example emission of greenhouse gasses). On the other hand, they contain significant amounts of sugars (both soluble and insoluble) and they can be used as raw material for the production of ethanol. In Chapter 3, Matsakas and colleagues utilized household food wastes as raw material for the production of ethanol at high dry material consistencies. A distinct liquefaction/saccharification step has been included to the process, which rapidly reduced the viscosity of the high solid content substrate, resulting in better mixing of the fermenting microorganism. This step had a positive effect in both ethanol production and productivity, leading to a significant increase in both values, which was up to 40.81% and 4.46 fold, respectively. Remaining solids (residue) after fermentation at 45% w/v dry material (which contained also the unhydrolyzed fraction of cellulose), were subjected to a hydrothermal pretreatment in order to be utilized as raw material for a subsequent ethanol fermentation. This led to an increase of 13.16% in the ethanol production levels achieving a final ethanol yield of 107.58 g/kg dry material. In conclusion, the ability of utilizing household food waste for the production of ethanol at elevated dry material content has been demonstrated. A separate liquefaction/saccharification process can increase both ethanol production and productivity. Finally, subsequent fermentation of the remaining solids could lead to an increase of the overall ethanol production yield.

In Chapter 4, Lam and colleagues studied the feasibility of using pastry waste as resource for glucoamylase (GA) production via solid state fermentation (SSF). The crude GA extract obtained was used for glucose production from mixed food waste. The results showed that pastry waste could be used as a sole substrate for GA production. A maximal GA activity of 76.1 ± 6.1 U/mL was obtained at Day 10. The optimal pH and reaction temperature for the crude GA extract for hydrolysis were pH 5.5 and 55 °C, respectively. Under this condition, the half-life of the GA extract was 315.0 minutes with a deactivation constant (kd) 2.20×10^{-3} minutes^{-1}. The application of the crude GA extract for mixed food waste hydrolysis and glucose production was successfully demonstrated. Approximately 53 g glucose was recovered from 100 g of mixed food waste in 1 h under the optimal digestion conditions, highlighting the potential of this approach as an alternative strategy for waste management and sustainable production of glucose applicable as carbon source in many biotechnological processes.

One of the most practical ways to utilise municipal solid waste is composting, thereby producing materials that may be productively used to improve soil properties. Wastes, as well as mature composts, contain hydrophobic substances, including fats, which are more resistant to microbiological decomposition than other constituents. The aim of Bekier and colleagues in Chapter 5 was to determine qualitative and quantitative changes of hydrophobic substances, especially fatty acids, during the course of municipal solid waste composting. This provides new information on intensity of hydrophobic versus other substances decomposition undergoing during these processes. Raw materials, prepared according to MUT-DANO technology, were composted in a pile, and samples were taken after 1, 14, 28, 42, 56, 90 and 180 days of the composting. Temperature, moisture, total organic carbon, hydrophobic substances carbon (HSC) and fatty acid carbon (FAC) contents were determined in all samples. Hydrophobic substances were extracted with 1:2 (v/v) mixture of ethanol/benzene, while fats were extracted with petroleum ether and determined by GC analysis after transesterification with BF_3 in methanol. The HSC decreased from 27.8 to 9.3 g kg^{-1} during first 90 days of composting, and thereafter remained constant. Similarly, the highest content of FAC was in raw compost, while the lowest was after 90 days. Octadecenoic acid predominated in the raw compost and decreased from 56 to 23 % FAC after 180 days. During the composting processes, domination of octadecenoic acid was replaced by hexadecanoic and octadecanoic acids, which increased from 18.8 to 36.7 % and 8.3 to 19.4 % FAC, respectively. The share of hexadecanoic, eicosanoic and docosanoic acids increased after the thermophilic phase. The presence of odd-numbered fatty acids (pentadecanoic and heptadecanoic) was noted, which are known to be products of the bacterial transformation-synthesis of lipid substances. The extent of decomposition of hydrophobic substances, especially fatty acids, is greater than other components in composted municipal solid waste, and intensity of the biotransformation is significantly correlated with composting parameters, mainly temperature and time. During the thermophilic phase of municipal solid waste composting, the decrease in total content of hydrophobic substances is approximately fivefold, while the reduction in fatty acids can be about tenfold. Unsaturated fatty acids are more intensively decomposed during the composting processes, while saturated fatty acids are more resistant. Moreover, transformation of fatty matter may result in the creation of specific isomers with odd numbers of carbon atoms.

Guam is a small, isolated tropical island in the western Pacific with a population of over 160,000 people. Although population growth and life style have been shown to have strong effects on the character and generation of waste, very little is known about consumption patterns and behavior of the people of Guam in this regard. Currently landfilling is the only discard method available to the island. Placement of huge volumes of organic waste material in landfills not only causes environmental problems for the island but in fact constitutes loss of valuable resources that could be composted and made available for land application as a soil amendment in forest lands, farm fields, and home gardens. Composting on the other hand reduces both the volume and the mass of the raw material while transforming it into a valuable soil conditioner. In Chapter 6, Golabi and colleagues present some of the results of survey questionnaires that was developed and conducted over the past two years that is anticipated to help waste operating managers and decision makers to determine societal consumption behavior and residential life style as the first step toward development of an effective waste-management strategy for the island of Guam. In this regards, they are also presenting an example of a large scale composting method developed in Isfahan, Iran, for recycling of organic wastes of municipal origin.

In Chapter 7, Resasco and colleagues present a review of the different conversion strategies for the catalytic upgrading of furfurals, specifically furfural and hydroxymethyl furfural, which are two chemical building blocks from lignocellulosic biomass for the production of transportation fuels, as well as useful acids, aldehydes, alcohols and amines. Reactions and catalysts for aldolcondensation, etherification, hydrogenation, decarbonylation and ring opening of furfurals are discussed. Specific examples are reviewed for hydrogenation and decarbonylation with emphasis on reaction pathways and kinetics analysis, comparing the behaviour of different metal catalysts.

The planning actions in municipal solid waste (MSW) management must follow strategies aimed at obtaining economies of scale. At the regional basin, a proper feasibility analysis of treatment and disposal plants should be based on the collection and analysis of data available on production rate and technological characteristics of waste. Considering the regulations constraint, the energy recovery is limited by the creation of small or medium-sized incineration plants, while separated collection strongly influences the heating value of the residual MSW. Moreover, separated collection of organic fraction in non-densely populated area is burdensome and difficult to manage. Chapter 8, by Trulli and colleagues, shows the results of the analysis carried out to evaluate the potential energy recovery using a combined cycle for the incineration of mechanically pre-treated MSW in Basilicata, a non-densely populated region in Southern Italy. In order to focalize the role of sieving as pre-treatment, the evaluation on the MSW sieved fraction heating value was presented. Co-generative (heat and power production) plant was compared to other MSW management solutions (e.g., direct landfilling), also considering the environmental impact in terms of greenhouse gases (GHGs) emissions.

Biomass and municipal solid waste offer sustainable sources of energy; for example to meet heat and electricity demand in the form of combined cooling, heat and power. Combustion of biomass has a lesser impact than solid fossil fuels (e.g. coal) upon gas pollutant emissions, whilst energy recovery from municipal solid waste is a beneficial component of an integrated, sustainable waste management programme. Concurrent combustion of these fuels using a fluidised bed combustor may be a successful method of overcoming some of the disadvantages of biomass (high fuel supply and distribution costs, combustion characteristics) and characteristics of municipal solid waste (heterogeneous content, conflict with materials recycling). It should be considered that combustion of municipal solid waste may be a financially attractive disposal route if a 'gate fee' value exists for accepting waste for combustion, which will reduce the net cost of utilising relatively more expensive biomass fuels. In Chapter 9, Laryea-Goldsmith and colleagues supressed emissions of nitrogen monoxide and sulphur dioxide for combustion of biomass after substitution of biomass for municipal solid waste materials as the input fuel mixture. Interactions between these and other pollutants such as hydrogen chloride, nitrous oxide and carbon monoxide indicate complex, competing reactions occur between intermediates of these compounds to determine final resultant emissions. Fluidised bed concurrent combustion is an appropriate technique to exploit biomass and municipal solid waste resources, without the use of fossil fuels. The addition of municipal solid waste to biomass combustion has the effect of reducing emissions of some gaseous pollutants.

Carbonization is a newly developed process that converts sewage sludge to biocoal, a type of solid biomass that can partially substitute for coal during power generation. Chapter 10, by Wang and colleagues, presents an assessment of the environmental effects of various sewage sludge treatment processes, including carbonization, direct landfills, co-incineration with municipal solid waste, and mono-incineration in Taiwan. This assessment was conducted using the life cycle assessment software SimaPro 7.2 and the IMPACT2002+ model. Results show that carbonization is the best approach for sewage sludge treatment, followed in descending order by co-incineration with municipal solid waste, direct landfills, and mono-incineration. The carbonization process has noticeable positive effects in the environmental impact categories of terrestrial ecotoxicity, aquatic ecotoxicity, land occupation, ionizing radiation, aquatic eutrophication, non-renewable energy, and mineral extraction. For the emission quantity of greenhouse gases, landfilling has the greatest impact (296.9 kg $CO2$ eq./t sludge), followed by mono-incineration (232.2 kg CO_2 eq./t sludge) and carbonization (146.1 kg CO_2 eq./t sludge). Co-incineration with municipal solid waste has the benefit of reducing green house gas emission (−15.4 kg CO_2 eq./t sludge). In the aspect of energy recovery, sewerage sludge that has been pretreated by thickening, digestion, and dewatering still retains a high moisture content, and thus requires a significant amount of energy use when used as a substitute solid fuel. Therefore, the carbonization of sewage sludge would be a more sustainable option if the energy delivery and integration processes are made more efficient.

Gasification is a thermo-chemical process to convert carbon-based products such as biomass and coal into a gas mixture known as synthetic gas or syngas. Various types of gasification methods exist, and fluidized bed gasification is one of them which is considered more efficient than others as fuel is fluidized in oxygen, steam or air. Chapter 11, by Begum and colleagues, presents an experimental and numerical investigation of fluidized bed gasification of solid waste (SW) (wood). The experimental measurement of syngas composition was done using a pilot scale gasifier. A numerical model was developed using Advanced System for Process ENgineering (Aspen) Plus software. Several Aspen Plus reactor blocks were used along with user defined FORTRAN and Excel code. The model was validated with experimental results. The study found very similar performance between simulation and experimental results, with a maximum variation of 3%. The validated model was used to study the effect of air-fuel and steam-fuel ratio on syngas composition. The model will be useful to predict the various operating parameters of a pilot scale SW gasification plant, such as temperature, pressure, air-fuel ratio and steam-fuel ratio. Therefore, the model can assist researchers, professionals and industries to identify optimized conditions for SW gasification.

The disposal of plastic solid waste (PSW) has become a major worldwide environmental problem. New sustainable processes have emerged, i.e. either advanced mechanical recycling of PSW as virgin or second grade plastic feedstock, or thermal treatments to recycle the waste as virgin monomer, as synthetic fuel gas, or as heat source (incineration with energy recovery). These processes avoid land filling, where the non-biodegradable plastics remain a lasting environmental burden. Within the thermal treatments, gasification and pyrolysis gain increased interest. Gasification has been widely studied and applied for biomass and coal, with results reported and published in literature. The application to the treatment of PSW is less documented. Gasification is commonly operated at high temperatures (>600°C to 800°C) in an airlean environment (or oxygen-deficient in some applications): the air factor is generally between 20% and 40% of the amount of air needed for the combustion of the PSW. Gasification produces mostly a gas phase and a solid residue (char and ashes). The use of air introduces N_2 in the product gases, thus considerably reducing the calorific value of the syngas, because of the dilution. Chapter 12, by Brems and colleagues, reviews the existing literature data on PSW gasification, both as the result of laboratory and pilot-scale research. Processes developed in the past are illustrated. Recently, the use of a sequential gasification and combustion system (at very high temperatures) has been applied to various plastic-containing wastes, with atmospheric emissions shown to be invariably below the legal limits. Operating results and conditions are reviewed in the paper, and completed with recent own lab-scale experimental results. These results demonstrate that gasification of PSW can be considered as a first order reaction, with values of the activation energy in the order of 187 to 289 kJ/mol as a function of the PSW nature.

PART I

FOUNDATIONS

CHAPTER 1

Energy Recovery from Municipal and Industrial Wastes: How Much Green?

SATINDER KAUR BRAR, SAURABH JYOTI SARMA,
AND MAUSAM VERMA

Waste feedstock, including and industrial wastes can be transformed into various forms of fuels that can be used to supply energy. The wasteto-energy technologies can be used to produce biogas (methane and carbon dioxide), syngas (hydrogen and carbon monoxide), liquid biofuels (ethanol and biodiesel), or pure hydrogen; and later, these fuels can then be converted into electricity. This transformation can be facilitated by various physical, thermal and biological methods. These processes have been driven by many technical drivers, such as the need for improved pollution and emissions controls for combustion, advanced non-incineration conversion methods, and hydrogen production enabling other clean technologies, such as fuel cells. Likewise, the strategic drivers, such as reduction in land filling, reduced dependence on fossil fuels, decreased greenhouse gas emissions and pollution and eligibility for carbon credits and tax incentives has been fueling the energy production from wastes. Despite the technical and strategic drivers, the energy recovery from waste often runs into dry owing to various technological bottlenecks, such as lack of ver-

Energy Recovery from Municipal and Industrial Wastes: How Much Green?. © Brar SK, Sarma SJ, and Verma M. Hydrology: Current Research S5 (2012), doi: 10.4172/2157-7587.S5-e001. Licensed under Creative Commons Attribution License, http://creativecommons.org/licenses/by/3.0.

satility (each system is specific for each type of waste); waste-gas cleanup and conversion efficiency (consuming more energy than producing it). In addition, there are strategic challenges, such as regulatory hurdles, high capital costs and opposition from environmental and citizen groups (social backlash).

In the existing world of mounting energy prices, population growth, and concerns regarding greenhouse-gas emissions, the need for alternative energy and alternatives to landfills and livestock waste lagoons has to increase. Further, bioethanol producers have begun to face the irk of their "environmentally friendly" products relying too heavily on fossil fuels for their production, and they are now using biogas from landfills or feedlots to power their refineries—biogas power for biofuels.

Among different thermal methods of waste management pyrolysis, gasification and combustion are the technologies commonly used for simultaneous waste management and energy recovery. Though these methods have been successfully used even in pilot scale, still they have certain environmental concerns. Similar to any other process used in waste management pyrolysis also has a few shortcomings which need to be considered for efficient/sustainable energy recovery using this technology. Firstly, the products (liquid/gaseous) of the process are complex. Secondly, it may use the wastes which are actually recyclable. Likewise, the process may utilize the organic part of the waste which otherwise could be used for other highly sustainable process such as composting. Further, requirement of high temperature could be another disadvantage of pyrolysis process. For example, a plasma pyrolysis vitrification process may require a temperature between 5000-14,000°C. Therefore, if the energy required to run the process is obtained from a sustainable source then only it may be considered as a green technology for energy recovery. Likewise, gasification of waste for energy recovery has also some issues regarding its sustainability. Firstly, the process may not have very high carbon sequestration efficiency as carbon dioxide may be released. Similarly, during the process toxic substances such as heavy metals and halogens could be released into the environment [1]. Combustion is another method for energy recovery by the utilization of waste. Presently, United States alone has nearly 86 plants for energy recovery by the combustion of municipal solid waste. However, combustion of waste materials for energy recovery has also certain serious concerns. Firstly, in terms of pollutant content, the gaseous emission of waste combustion process is almost similar to energy recovery by fossil fuel combustion. Likewise, the process needs proper management of the ash (fly ash or bottom ash) generated during the process. Additionally, possible release of heavy metals and polyaromatic hydrocarbons during energy recovery by combustion of waste material is another issue [2].

As already mentioned, waste materials could also be subjected to biological processes for energy recovery. Among different bio-based techniques anaerobic digestion is comparatively simple, common and old process. However, it has a range of disadvantages/technical constraints which need to be assessed and resolved to make this technology one of the most efficient technology for energy recovery. Firstly, similar to any other process for the production of gaseous fuel it has a risk of fire and explosion. Likewise, the cost associated with collection, transportation as well as processing of the waste materials may be prohibitory for economic feasibility of the process. Further, harmful emission due to transportation of the waste materials and operation of the process could be considerable with respect to the environmental benefit of the process. Moreover, the efficiency of the process is directly dependent on the organic content of the waste feedstock and the waste with less organic material is not suitable for the process. Hence, waste separation may be required to improve the energy conversion efficiency of the system. Additionally, presence and propagation of pathogen in the putrescible substrate is another serious concern of anaerobic digestion process. Further, according to a report from US-EPA the process of energy recovery by anaerobic digestion of waste is not an economically feasible process until it is integrated to another source of revenue [3].

Besides anaerobic digestion and thermal methods, landfill gas recovery is another successful method of energy recovery from waste materials. Landfill gas is mainly composed of methane (and CO_2) which could be considered as the most dangerous greenhouse gas as it is nearly 20 times potent than CO_2. Methane emission from landfill sites is a serious problem as landfills are one of the three major sources (16% of total methane emission) of methane emission in United States [4]. Therefore, considering the problem of global warming, utilization of landfill gas as a source of renewable energy is recognized to have an additional benefit of greenhouse gas reduction. However, often the techniques used to recover the landfill gas can capture only a small fraction of methane produced in a landfill and the release of methane during landfill gas recovery process could have serious environmental consequences [5]. Moreover, due to high capital cost involved, landfill gas recovery may not be economically feasible for smaller landfill sites.

Thus, apart from the concerns associated with techno-economic feasibility energy recovery from wastes, they have certain environmental issues which need to be seriously dealt with prior to its long term application. In retrospective, energy recovery from municipal and industrial wastes can be often energy intensive and thus, theoretically would turn out to be a non-environmental option. However, taking into consideration the value-added effects of the energy recovery, such as simultaneous treatment and detoxification, the energy recovery is globally a green approach.

REFERENCES

1. Salman Zafar (2009) Gasification of municipal solid waste.
2. Methods of waste incineration.
3. (2005) Anaerobic Digestion: Benefits for Waste Management, Agriculture, Energy, and the Environment. Environmental protection agency.
4. Greenhouse Gas Emissions, United states environmental protection agency.
5. The DANGER of Corporate Landfill Gas-to-Energy Schemes and How to Fix It.

CHAPTER 2

Energy Recovery Potential and Life Cycle Impact Assessment of Municipal Solid Waste Management Technologies in Asian Countries Using ELP Model

ANDANTE HADI PANDYASWARGO, HIROSHI ONODA, AND KATSUYA NAGATA

2.1 BACKGROUND

Developing countries in Asia have a number of similarities in terms of their waste composition and characteristics. High moisture content due to the high percentage of organic waste composition results in low calorific value. This makes it is less suitable for thermal treatments and more suitable for biological treatments, such as composting and anaerobic digestion. The organic fraction of municipal waste equates to 62% in Indonesia, 63.4% in China, and 41.8% in India [1]. Waste volume in this region increases with the growth of population, urbanization, industrialization,

---

Energy Recovery Potential and Life Cycle Impact Assessment of Municipal Solid Waste Management Technologies in Asian Countries Using ELP Model. © Pandyaswargo AH, Onoda H, and Nagata K. International Journal of Energy and Environmental Engineering 3,28 (2012). doi:10.1186/2251-6832-3-28. *This article was originally distributed under a Creative Commons Attribution 2.0 Generic License, http://creativecommons.org/licenses/by/2.0/.*

and economic development. Indonesia, with a population of 232.7 million, generates 38.5 million tonnes of municipal waste annually. China, with a population of 1.3 billion, generates 1.8 billion tonnes. India, with a population of 1.2 billion, generates 66.69 million tonnes [1]. Another important waste characteristic is the generation rate per capita per day, which is 0.75 kg in Indonesia, 1.54 kg in China, and 0.2 to 0.5 kg in India, depending on the size of the city. Figures 1 and 2 summarize the waste characteristics in the three countries.

TABLE 1: Current trends of municipal waste treatment technologies applied in the countries

Country	Landfill	Anaerobic digestion	Composting facility	RDF	Incineration facility	Other
Indonesia	68.86% is landfilled; only 10% of the landfills are sanitary landfill	NA	7.19% of the municipal waste is composted	NA	4.49% are burned in the open space, 6.59% are burned in small scale incineration plant	2.99% of the waste are dumped into the river, 9.58% are buried
China	56.6% of waste dumped is into the landfill, and 28.6% are open-dumped	NA	12.9% of waste is composted	NA	1.9% of waste is incinerated	
India	Non-existence of sanitary landfill; open dumping is common	Unsuccessful large-scale AD plants in Nagpur, Lucknow, Vijaywada (20 t/day), and Koyambedu flower market (30 t/day) due to low-quality input	Vermi-composting and aerobic windrow composting are practiced in clusters; product quality is not optimal	Unsuccessful RDF plants in Deonar, Mumbai(80 t/day), Bangalore (5 t/day), Hyderabad (700 t/ day), and Vijaywada (600 t/ day) due to low calorific value	Unsuccessful incineration plant in Timarpur (300 t/day). Two on- trial incineration plants in Delhi (1,950 t/day) and (1,300 t/day)	

Source: [1]. RDF, refuse-derived fuel.

Energy Recovery Potential and Life Cycle Impact Assessment

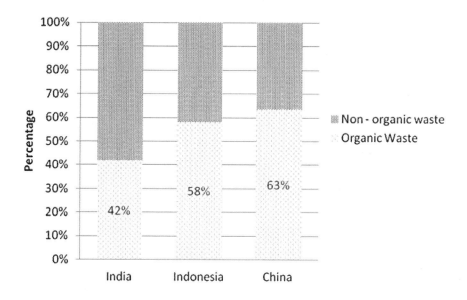

FIGURE 1: Municipal organic waste percentage in India, Indonesia, and China.

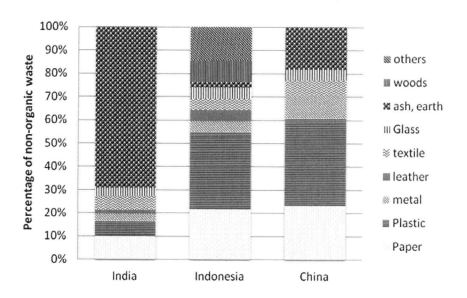

FIGURE 2: Municipal non-organic waste compositions in India, Indonesia, and China.

The current common practice of municipal waste treatment in these countries is the use of landfill. Almost 68.86% of waste generated in Indonesia is landfilled, and only 10% of the available landfills meet the requirements for a sanitary landfill. China deposits 56.6% of its municipal waste in landfills. India has a very limited number of sanitary landfills, and thus, open dumping is common and widespread. Incineration is not applied on a large scale in Indonesia, whereas 1.9% of China's municipal waste is incinerated. India initiated several projects to incinerate its waste, such as the 300 t/day capacity incineration plant in Timarpur. However, such projects proved to be unsuccessful [2]. Two pilot projects on incineration plants are ongoing in Delhi, with 1,950 and 1,300 t/day capacities [1]. In Indonesia, composting of municipal waste is done both on the community level and on the final disposal site and utilizes 7.19% of the waste generated in Indonesia. A respectable 12.9% of waste is composted in China [1]. In India, vermi-composting and windrow composting are practiced in clusters. The municipal waste compost fertilizers often have difficulties in competing with the chemical fertilizers due to lower nutrient content and the presence of heavy metals.

Anaerobic digestion (AD) has proven to be quite successful when applied on a small scale but rather unsuccessful when applied on a large scale. For example, the 20 t/day and 30 t/day AD plants in Nagpur, Lucknow, Vijaywada, and Koyambedu flower market in India failed due to low-quality input [1,3]. Table 1 shows the waste treatment trend in each of the countries in this study.

The main objective of this study is to identify the most appropriate technology to be adopted in the region by using the environmental load point (ELP) methodological approach. Furthermore, this study attempts to measure the environmental impact and energy recovery potential by applying life cycle assessment (LCA) to different scenarios of waste treatment in the three countries. The scenarios proposed are adjusted to the waste characteristics, which can be described as mixed waste with high organic composition. Composting and anaerobic digestion are proposed to favor the high organic composition, incineration is proposed to respond to the mixed state of the waste collected, and landfill gas collection for energy recovery is chosen to make use of the sanitary landfill that already exists in the region.

2.2 METHODS

2.2.1 METHODOLOGICAL APPROACH

2.2.1.1 STUDY AREAS

Three developing countries in Asia—Indonesia, China, and India—are selected as the study areas to represent Asia because of their similarities in high organic waste composition and the current low levels of energy and resource potential utilization.

2.2.1.2 DATA USED

Both primary and secondary data were used in this study. Primary data was inventory data collected from the EcoInvent database 2010 and Japan Environmental Management Association for Industry (JEMAI). Secondary data was collected from existing literature on emissions and energy recovery potentials of treatment processes. In addition, a questionnaire survey was conducted at national universities in the selected countries. The data collected was used to weigh the results in order to get a better representation of geographic, social, and political interests in the country [4].

2.2.1.3 METHODOLOGY USED

In the year 2002, a thorough guideline of LCA application in municipal solid waste (MSW) was prepared by Nordtest Finland [5]. This guideline, along with ISO14040 [6], ISO 14044 [7], and an LCA methodology study by Finnveden [8], has significant contributions in applying ELP methodology in this study. The ELP methodology is an Excel-based LCA tool that allows for a high degree of adjustability and transparency as well as social factor integration to refine and personalize results. This methodology has been used to assess municipal incinerators and water supply plants as well as product manufacturing factories in Japan [9-12]. The study done by Onoda analyzed six options for municipal waste management in Kitakyushu City. The business of usual incineration + ash landfilling was compared to five other scenarios elaborated in Table 2. The result showed that the scenario where the non-organic waste is incinerated and organic waste is digested anaerobically has the lowest Environmental Load Point, highest energy recovery, and lowest CO_2 emission.

TABLE 2: Previous ELP study result on Kitakyushu municipal waste management scenarios

Case	Scenario	ELP	Energy recovery	CO_2 emission
1	Incineration (electric) + ash landfilling	100	100	100
2	Incineration (electric) + ash melting + ash landfilling	93	100	102
3	Incineration + ash melting + metal recycling	92	100	102
4	Direct melting (gas) + ash landfilling	98	95	103
5	Direct melting (gas) + metal recycling	96	95	104
6	Incineration (electric) + organic waste anaerobic digestion	91	125	99

Since the processes are similar, the figures were adjusted relative to business as usual (case 1) as baseline unit (100). Source: [11].

The ELP has nine impact categories. These categories are energy depletion, ozone depletion, acid rain, resource scarcity, air pollution, ocean and water pollution, problem of waste disposal, and ecosystem effect. Each of the impact categories has indicators, such as oil, natural gas, and coal for energy depletion; CH_4 and CO_2 for global warming; chlorofluorocarbon and hydrofluorocarbon for ozone depletion; NOx and SOx for acid rain; Fe, Ni, Sn, Au, and Ag for resource scarcity; PM_{10} and $PM_{2.5}$ for air pollution; biological oxygen demand (BOD), chemical oxygen demand (COD), and suspended solids for ocean and water pollution; solid waste for problem of waste disposal; and petrol, benzene, and dioxin for ecosystem effect. The complete list of indicators is comprised of 186 items. Table 3 shows the impact categories and the indicators relevant to this study that were incorporated for analysis.

TABLE 3: ELP impact categories

Impact category (j)	Indicators (k)	Weight coefficient (C)[a]
Energy depletion	Oil, natural gas, coal	0.089
Global warming	CO_2, CH_4	0.082
Ozone depletion	(not used in this study)	0.098
Acid rain	NOx, SOx	0.086
Air pollution	SO_2, NO_2, CO, $PM_{2.5}$, PM_{10}	0.072
Resource consumption	Iron (Fe), nickel (Ni), tin (Sn), aluminum (Al), gold (Au), silver (Ag)	0.134
Ocean and water pollution	BOD, COD	0.135
Problem of waste disposal	Slag, residues	0.107
Ecosystem influence	(Not used in this study)	0.197

[a]*Fixed by laboratory and literature result [13].*

The mathematical equation for the LCA model consists of a three-step calculation. The first step is determining the annual load (A). A is the result of multiplying the weight coefficient of each item in the impact category fixed by the laboratory literature (C) listed in Table 3 with the national annual consumption and emission (TQ).

Equation 1. Annual load formula

$$A_j = \sum_k (C_{j,k} \times TQ_k) \tag{1}$$

- A_j, annual load in j impact category
- $C_{j,k}$, weight coefficient for k indicator in j impact category
- TQ_k, annual consumption or emission for k item
- Suffix j, impact category
- Suffix k, indicator in impact category

The second step is to calculate the environmental load factor (ELF), which is the result of multiplying the coefficient (C) listed in Table 3 and the weighting value (W), divided by the annual load (A) from the Equation 1 results. The weighting value (W) is derived from surveys and questionnaires of the related stakeholders or communities. In this study, the methodology of analytic hierarchical process (AHP) questionnaire was adopted.

Equation 2. Environmental load factor formula

$$ELF_k = \sum_j \left(C_{j,k} \times \frac{W_j}{A_j} \right) \qquad (2)$$

- ELF_k, integrated coefficient for k item
- W_j, weight coefficient (category importance) from questionnaire in j impact category
- $C_{j,k}$, weight coefficient for k indicator in j impact category
- A_j, annual load in j impact category
- Suffix k, indicator in impact category
- Suffix j, impact category

The final step is multiplying the ELF with the total indicator's consumption or emission of the process or production of the related MSW technologies from the EcoInvent database to get the environmental load point as the final output. Equation 3 shows the formula used for this calculation.

Equation 3. Environmental load point formula

$$ELP_i = \sum_k \left(ELF_k \times Q_{i,k} \right) \qquad (3)$$

- ELP_i, integrated indicator
- ELF_k, integrated coefficient for k indicator
- $Q_{i,k}$, total consumption or emission for k indicator in process i
- Suffix i, process or product
- Suffix k, indicator in impact category

2.2.2 LIFE CYCLE AND IMPACT ASSESSMENT

2.2.2.1 GOAL AND SCOPE DEFINITION

To find the most appropriate municipal solid waste management (MSWM) technology by using the ELP methodology, this study selected the most relevant six categories out of the nine available ELP impact categories. The six categories covered in this study are energy depletion, global warming, acid rain, resource consumption, air pollution, and waste disposal. The functional unit of the LCA study has been set as thousand metric tonnes. The incorporated inventory data are the resources taken from nature and the emissions released into the air and soil. The indicators taken into account in this study are oil, natural gas, coal, CO_2, CH_4, NOx, SO_2, Fe, Ni, Sn, Al, Au, Ag, SO_2, CO, $PM_{2.5}$, PM_{10}, slag, and residue from the waste management processes. The emission factors of conventional electricity production are taken from the electricity grid fuel mix of the country. The study estimated the emission from both the mining process and the power plants of fossil fuel-based electricity substituted by the energy recovered from the waste treatment processes. The study also estimated the net energy recovery potential of waste treated in each country as well as the emission from mineral fertilizer production substituted by the compost fertilizer.

2.2.2.2 SYSTEM BOUNDARY

The transport distance of waste from the sources to all process systems is assumed to be equal, and therefore, it is excluded from the system boundary. It is also assumed that the heat recovery, although identified in kWh-heat, is not usable due to the lack of a district heating system in the region. Due to the absence of a district heating system in the majority of the area in the region, the calculated heat recovery is not considered in the environmental impact avoided. However, further study on a cooling system by a heat exchanger that utilizes the heat waste would be recommendable. The other assumption is that the inventory data for each process are similar to those of the European and the Japanese databases available, which is why the use of local input data, such as waste composition, national emission and resource consumption, and the survey and questionnaire, provides significant contributions in personalizing the results of this study.

2.2.2.3 INVENTORY DATA ANALYSIS

Inventory data for the processes involved in the study were taken from the EcoInvent database 2010 [14], the JEMAI, and from literature that examined the existing local processes. The three scenarios constructed for the MSW treatment are a mix of incineration, sanitary landfilling, composting, and anaerobic digestion. In scenario 1, the entire amount of the mixed municipal waste is incinerated. In scenario 2, the organic waste is composted whereas the rest of the waste is landfilled in the sanitary landfill. The CO_2 and CH_4 (biogas) from the sanitary landfill are collected for energy recovery with a cogeneration unit. In scenario 3, the organic waste is digested anaerobically, and the rest of the waste is to be landfilled in the sanitary landfill, with the biogas emitted being collected for energy recovery. The desired output of the first scenario is electricity and heat. The desired output of the second scenario is electricity and compost fertilizer. The desired output of the third scenario is digested matter, which can be used as soil conditioner, and biogas to generate electricity. To estimate the emission avoided from fossil fuel-based electricity, the mining and electricity production from coal, natural gas, hydropower, and crude oil are accounted. Table 4 summarizes the scenarios and the desired output. The system boundary of each scenario is elaborated in Figures 3, 4, and 5.

TABLE 4: Scenario and output

Case	Scenario	Desirable output
1	Incineration + energy recovery	Electricity, heat
2	Composting + sanitary landfilling + landfill gas collection for energy recovery	Fertilizer, electricity, heat
3	Biogas + sanitary landfilling + landfill gas collection for energy recovery	Digested matter, electricity, heat

Energy Recovery Potential and Life Cycle Impact Assessment

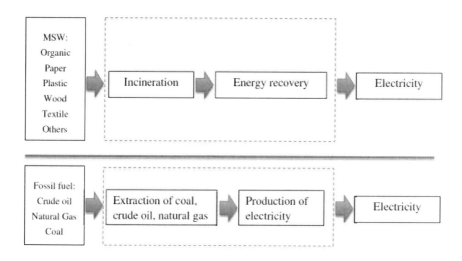

FIGURE 3: Scenario 1. Incineration + energy recovery (above the line) and the replaced process to produce the same amount of energy with the business as usual practice (below the line).

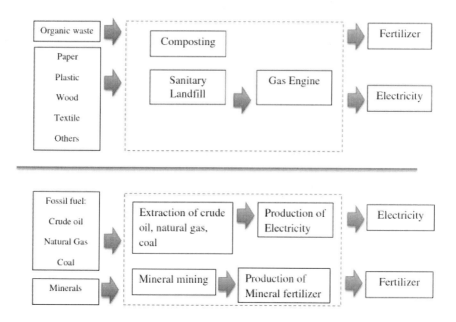

FIGURE 4: Scenario 2. Composting + sanitary landfill + energy recovery (above the line) and the replaced processes to produce the same amount of output with business as usual (below the line).

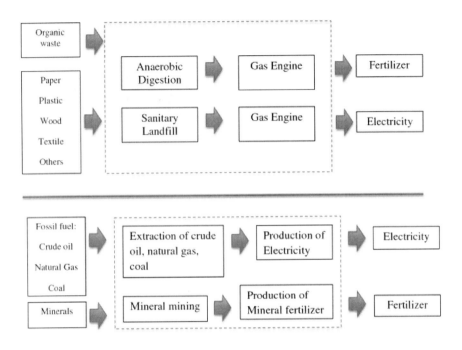

FIGURE 5: Scenario 3. Anaerobic digestion + sanitary landfill + energy recovery (above the line) and the replaced processes to produce the same amount of output with business as usual (below the line).

2.2.3 MSW TREATMENT TECHNOLOGIES

MSW may be treated thermally or biologically. Thermal treatment includes incineration, pyrolysis, and gasification, while biological treatment includes anaerobic digestion and composting. Thermal treatment requires high calorific value. Therefore, dry combustible waste such as plastic, rubber, and paper are desirable for this treatment. Biological treatment requires high organic content. Therefore, food waste and garden waste are desirable for this treatment methodology [15]. In this study, the technologies adopted in the scenario constructed are (1) sanitary landfill, (2) incineration, (3) composting, and (4) anaerobic digestion.

2.2.3.1 SANITARY LANDFILL

Landfill is still the common practice of MSWM in the developing world. Sanitary landfills, although quite limited in number, exist especially in larger cities. A sanitary landfill has a proper leachate capture system and liners to prevent contamination of the groundwater. Although landfill is a less preferable solution, especially for non-inert waste, due to the limited lifetime (30 to 50 years) and slow biodegradation process for organic waste [16], sanitary landfill is selected in this study as an option because of the possibility of landfill gas collection for energy recovery. Sanitary landfill inventory data used in this study include landfill gas incineration and landfill leachate treatment in the wastewater treatment plant (WWTP) as well as the WWTP sludge disposal in the municipal incinerator.

2.2.3.2 INCINERATION

Incineration is perceived as a costly solution for MSWM due to its operational energy requirements and the flue gas treatment. It is also technically feasible only for a relatively high calorific value of 1,433 kcal [17], which is often quite high for developing Asian countries' waste to meet. For example, the calorific value of Indian waste is only 700 to 1,000 kcal [18]. However, modern incinerators have improved with efficient combustors and flue gas treatments [16]. Moreover, some plants add auxiliary fuels like crop waste and/or tires to improve the calorific value. Significant amounts of methane gas released into the atmosphere are not achieved with this technique, especially when compared to the landfilling option. The inventory data used in this study for incineration include the landfilling of the residual materials, such as the fly ash and the scrubber sludge. The energy recovery potential per kilogram of waste incinerated is elaborated in Table 5.

2.2.3.3 COMPOSTING

Composting organic biodegradable waste takes a significant amount out of the waste stream going to incineration and landfill. This implies less landfill gas and leachate production. The bigger-scale composting plants in developing Asian countries often use open windrow composting. This aerobic composting approach typically takes about 4 to 6 weeks to reach the stabilized end-product stage. The composting process in this study incorporates emissions both from the energy demand for plant operation and infrastructure. The assumed water content is 50% by weight [15]. The assumed replaced mineral fertilizer is potassium nitrate (KNO3), as N. This mineral fertilizer has N content of 14%, while that of the municipal waste compost fertilizer ranges from 10% to 22% [19,20]. The release of N from mineral fertilizer is, however, quite significant in the first year (up to 80%) and low in the following years, while municipal waste compost fertilizer releases N gradually throughout the years (about 10% per year) [5]. Therefore, the amount of replaced mineral fertilizer is assumed to be equal.

TABLE 5: Electricity generated from waste incineration energy recovery

Type of waste	Net electricity produced per kilogram of waste treated (kWh/kg)
Biowaste	0.04
Paper	0.36
Plastic	0.96
Glass	0
Wood	0.36
Textiles	0.37
Others (20% water content)	0.28

Source: EcoInvent database, 2010.

2.2.3.4 ANAEROBIC DIGESTION

Anaerobic digestion is by far more efficient when compared to collecting landfill gas as the waste is processed in a closed container with conditioned temperature and the absence of oxygen creates the optimal environment for biogas generation. A study shows that a ton of waste in a controlled anaerobic digestion produces two to four times more methane in 3 weeks than a ton of waste in a landfill would produce in 6 to 7 years [21]. The input of anaerobic digestion should contain relatively pure organic material, the output being biogas with 55% to 60% CH_4 and 40% to 45% CO_2 that can be burned in a gas engine to generate electricity, and the residue being in the form of digested matters which can be used as soil conditioner. While biogas contains both CH_4 and CO_2, only CH_4 is considered to be convertible to electricity. Additionally, the heat value assumed in this study is 6 kWh/m³ CH_4[5]. The assumed digested matter usable as soil conditioner (fertilizer) in this study is 40% of the organic matter input. Spreading the product fertilizer from this process might take more energy when compared to mineral fertilizer because the nutrient content is less; thus, a larger amount is required. For this reason, the emission from the spreading activity is included.

2.2.3.5 REPLACED FOSSIL-BASED ELECTRICITY

The substituted electricity uses the national electricity grid thermal fuel mix. For example, in Indonesia, the fuel mix for JAMALI (Java, Madura, Bali) is used. This grid that provides 78% of the national electricity consumption [22] utilizes 53.7% natural gas, 18.74% coal, and 27.69% oil [23]. For China, the consumption of coal, oil, and gas of a thermal power plant is 96.62%, 1.87%, and 1.51%, respectively [24]. For India, the coal-based thermal power plant air emission [25] and the fuel mix of the thermal power plant, which is 82% coal, 17% gas, and 1% diesel [26], were used.

2.3 RESULTS AND DISCUSSION

The result of the first equation (A) is summarized in Table 6. The TQ value required for this calculation was collected from the government and institutions that provide the national annual consumption and emission of the related country, such as the US Energy Administration for the energy consumption [27], the Indonesian Ministry of Environment for the greenhouse gas (GHG) emission of Indonesia [28], and the United Nations Statistics [29], a study of air pollution in Asia [30], mining product consumption information from the National Statistics Office [31], China Mining Association [32], and index mundi [33].

TABLE 6: Annual load results derived by equation 1.

Impact category	India	Indonesia	China
Energy depletion	6.46E+10	1.49E+12	2.58E+12
Global warming	1.86E+13	1.00E+11	1.21E+14
Acid rain	9.00E+09	1.80E+09	4.56E+10
Resource consumption	2.46E+11	1.80E+09	1.62E+11
Air pollution	2.10E+10	2.10E+10	1.00E+11
Waste disposal	4.20E+10	4.20E+10	1.80E+09

To get the W value for the second step of the calculation using Equation 2, AHP questionnaires were distributed. Respondents are randomly selected from faculties in top universities in the related countries, such as the Institute of Technology Bandung, Indonesia, University of Delhi, India, and Beijing University, China. University students were selected as group of respondents for the ease of regular updating and comparability across countries. Figure 6 shows the questionnaire results. In the questionnaire, respondents were asked to compare and rate which of the nine ELP impact categories deserve the priority of concern in their countries and which deserve less. According to the total 300 university students surveyed in the three countries, energy depletion comes in the first rank of the most important impact category in Indonesia and China, while global warming is the most important issue in India. On the second rank is global warming in Indonesia, resource consumption in China, and ozone depletion in India.

Table 7 summarizes the ELF result. ELF is the value of ELP per kilogram emission or resources emitted or consumed in a process. Figure 7 summarizes the total of ELP quantification results of the three scenarios constructed in each country. The description of the results is described country-wise for each impact category, followed with a logical discussion in order to support the results.

Energy Recovery Potential and Life Cycle Impact Assessment

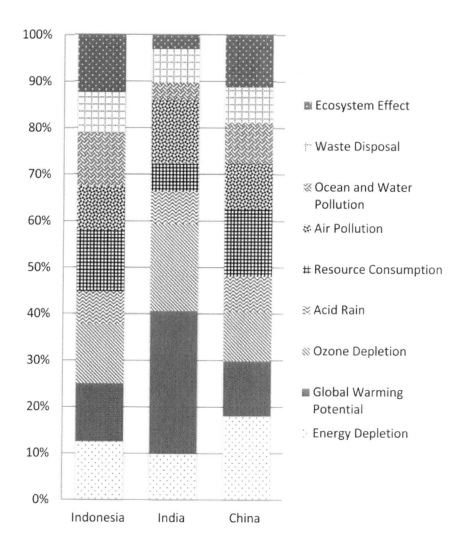

FIGURE 6: Weighting values from the AHP questionnaire.

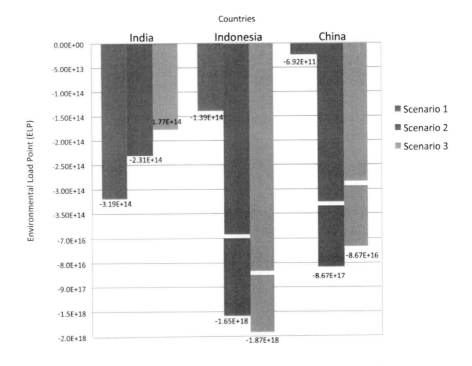

FIGURE 7: Total ELP of scenarios 1, 2, and 3 in India, Indonesia, and China.

TABLE 7: ELF results derived by equation 2

Impact category	India	Indonesia	China
Energy depletion	3.65E+09	8.25E+06	2.55E+07
Global warming	1.52E+05	1.02E+06	1.02E+06
Acid rain	2.94E+05	9.36E+05	9.36E+05
Resource consumption	3.25E+02	4.59E+09	4.59E+09
Air pollution	1.31E+06	3.24E+07	3.24E+07
Waste disposal	1.70E+04	2.37E+04	2.37E+04

The major findings extracted through our analysis are following:

2.3.1 INDIA

The environmental load of the 'energy depletion' impact category in the Indian case study is lowest in scenario 1, second lowest in scenario 3, and highest in scenario 2. This is because electricity replaced by energy recovered in the incineration plant significantly reduced the consumption of coal and natural gas in the fossil-based fuel thermal power plant. The 'global warming potential' impact category is lowest in scenarios 2 and 3. This is mainly due to the biological processes in these scenarios which take out CH_4 from the global warming potential, as well as the subsequent conversion of this gas into electricity. The 'acid rain' impact category, which consists of NOx and SO_2 as indicators, has the lowest impact in scenario 2, second lowest in scenario 3, and highest in scenario 1. The biggest contribution is from the avoided NOx emission from the production of mineral fertilizer. The 'resource consumption' impact category, which has Fe, Ni, Sn, Al_2O_3, Au, and Ag as indicators, has the lowest environmental load in scenario 3, followed by scenarios 2 and 1, as the amount of fertilizer produced by anaerobic digestion replaces the production of mineral fertilizer. The 'air pollution' impact category has the lowest environmental load in scenario 1, followed by scenarios 2 and 3, especially because of the NOx emission from the biogas and landfill gas cogeneration units. The 'waste disposal' impact category is highest in scenario 1 because the amount of inert material contained in Indian municipal waste produces significant amounts of slag and residues.

From the three scenarios proposed in the Indian study, the lowest environmental load is from scenario 1, in which the whole volume of municipal waste is incinerated. This result is mainly contributed by the significant amount of coal (82%) used in the Indian electricity grid fuel mix. Moreover, the weighting from community survey by the AHP questionnaire in this study ranked global warming potential and air pollution in the top three most concerning environmental issues in India. The estimated net electricity generated from combusting 1,000 tonnes of Indian waste in the incineration plant is 208 MWh.

In reality, incineration is not feasible for Indian waste due to its low calorific value. The refuse-derived fuel (RDF) method, which increases the calorific value of waste by taking out the moisture content by gasification and pelletization before feeding it to the incineration plant, is practiced. The product of RDF is often mixed into the coal power plant [1].

2.3.2 INDONESIA

The environmental load of the 'energy depletion' impact category in the Indonesian case study is lowest in scenario 3, mainly due to the avoided energy to produce the mineral fertilizer replaced by the digested matter from anaerobic digestion. The 'global warming potential' impact category, whose indicators are CH_4 and CO_2, is lowest in scenario 3, mainly because of the closed tank of anaerobic digestion preventing gas release into the atmosphere and enabling its conversion into electricity. The 'acid rain' impact category is lowest in scenarios 2 and 3, mainly because of the emission avoided from the production of replaced mineral fertilizer. Similarly, the lowest environmental load for the 'resource consumption' impact category also lies in scenarios 2 and 3 because of the resources saved from the replaced mineral fertilizer production. The 'air pollution' impact category is highest in scenario 1 and lowest in scenario 3, mainly because of the CO and NOx emitted by the incineration plant. The 'waste disposal' impact category is highest in scenario 2 due to the amount of waste going to the landfill plus the residual waste from the composting activity.

Energy Recovery Potential and Life Cycle Impact Assessment

Among the three scenarios in the Indonesian case study, scenario 3, which is anaerobic digestion for the organic waste content and landfill gas collection for energy recovery, has the least environmental load. This is mainly contributed by the digested matter replacing mineral fertilizer. The fertilizer produced is a co-benefit of anaerobic digestion. This means that no additional input of energy or resources is required to produce fertilizer, and all of the potential energy is captured within the closed container of the biogas plant. Moreover, the Indonesian survey results for weighting rank resource consumption as the most important impact category. The estimated electricity recovered from anaerobic digestion in the Indonesian waste case study is 32.7 MWh for every 580 tonnes of organic waste treated, and the estimated electricity recovered from landfill gas collection is 57.7 MWh for every 370 tonnes of non-inert, non-biowaste dumped in the sanitary landfill. The estimated amount of digested matter for soil conditioner is 232 tonnes for every 580 tonnes of organic waste treated in the anaerobic digestion plant.

In practice, the technology of large-scale municipal waste aerobic digestion is not popular in Indonesia [34]. This technology is commonly applied to animal slurry or agricultural waste because of the pure organic waste content. However, countries like the Netherlands, Sweden, and Switzerland have fully developed anaerobic digestion plants for handling municipal waste [15].

2.3.3 CHINA

The environmental load of the 'energy depletion' impact category in the Chinese case study is lowest in scenario 2, mainly due to the avoided coal and oil for mineral fertilizer production. The 'global warming' impact category is also lowest in scenario 2, mainly because of the avoided CO_2 emission from the production of the replaced mineral fertilizer. The lowest environmental load in the 'acid rain' impact category is in scenario 3 because of the avoided NOx and SOx emission from the replaced mineral fertilizer production. In the 'resource consumption' impact category, scenario 3 has the lowest environmental load. The biggest contribution to the load reduction is from the avoided iron and nickel consumed in the production of replaced mineral fertilizer. The impact category of 'air pollution' has the lowest environmental load in scenario 2. The $PM_{2.5}$ and NOx emission avoided from the replaced mineral fertilizer have the biggest contribution to this result. Finally, the highest environmental load in the 'waste disposal' impact category is scenario 2 because of the higher amount of waste composted, resulting in a higher amount of residual waste from the composting activity.

Among these three scenarios in the Chinese case study, scenario 2, which is composting for organic waste and sanitary landfill with energy recovery, has the lowest environmental load point for the Chinese case study. This is mainly due to the large percentage of organic waste (63%) within the Chinese waste composition and the weighting of resource consumption as being the second most important impact category. Moreover, the impact category of waste disposal is scored as the lowest weight by the Chinese survey respondents. This makes the volume of waste dumped in the sanitary landfill less significant.

In practice, large-scale composting is practiced in large Chinese cities such as Beijing, Shanghai, and Urumqi. These plants are often registered as Clean Development Mechanism projects, receiving carbon credits [1]. The estimated compost fertilizer produced in the Chinese second scenario case study is 254 tonnes of waste for every 630 tonnes of waste treated. The recovered energy from landfill gas collection is 60 MWh for every 246 tonnes of non-inert, non-biowaste dumped in the sanitary landfill.

Considering the salient results, this study suggests scenario 2 (composting) as the preferable option because of its appearance as the second best option in two of the countries (India and Indonesia) and as the best option for China. This means that the environmental impact and risk of failure are considered as medium compared to the other options. This scenario also offers a significant amount of energy recovery potential from the sanitary landfill.

2.4 CONCLUSIONS

This study tried to analyze the LCA impact assessment using an ELP model for the very first time in the three selected Asian countries. Each goal that was set at the start of the study was achieved, and it was concluded that each country in this study gave different resulting scenarios of the lowest environmental load. The best result for the Indian case is the scenario with incineration; for the Indonesian case, the best result is the scenario with anaerobic digestion, and the best Chinese result is the scenario with composting. These are a reflection of what is most valued in the country based on the university students' questionnaire results, the waste characteristics, the country's annual emission and consumption, as well as the emission and energy recovered by each of the technology options. The unique and different results for each country show that the ELP methodology has the potential to provide personalized results that incorporate technical and social considerations.

This study was a preliminary effort to apply this methodology, and the results are only theoretical estimations. Future research in this field could enhance the results of this study by including economic and social evaluations of each scenario or by increasing the scope of countries to obtain a larger data field for validation.

REFERENCES

1. IGES: National overview/country situation report on municipal solid waste management and composting; interim reports, Institute of Global Environmental Strategies, Kitakyushu (2012)
2. Sharholy, M, Ahmad, K, Mahmood, G, Trivedi, RC: Municipal solid waste management in Indian cities - a review. Waste Manag.. 28, 459–467 (2008).
3. Klöpfer, W: Life cycle assessment from the beginning to the current state. Environ. Sci. Pollut. Res. . 4, 223–228 (1997).
4. Pandyaswargo, AH, Onoda, H, Nagata, K: Integrated LCA for solid waste management in developing country. Proc. EcoDes. 2011 Int. Symp. (2011).
5. Bjarnadottir, HJ, Frioriksson, GB, Johnsen, T, Sletsen, H: Guidelines for the use of LCA in the waste management sector, Nordtest, Finland (2003)
6. ISO 14040: Environmental management - life cycle assessment - principles and framework. ISO 14040:2006, International Organization for Standardization, Switzerland (2006)
7. ISO 14044: Environmental management - life cycle assessment - requirements and guidelines. ISO 14044:2006, International Organization for Standardization, Switzerland (2006)
8. Finnveden, G: Methodological aspects of life cycle assessment of integrated solid waste management systems. Resour. Conserv. Recycl.. 26, 173–187 (1999).

9. Nagata, K, Nohtomi, M, Sekiai, H, Okachi, T, Ohashi, Y, Osada, M: Study on the environment load assessment techniques for the effective use of water resources. Soc. Mech. Eng.. 3, 406–409 (2003)
10. Hu, H, Onoda, H, Nagata, K: Life cycle assessment of municipal waste management processes based on real data. J. Waste Manage. Assoc.. 64, 83–89 (2011) Available in Japanese
11. Onoda, H: Life cycle assessment of incineration ash recycling and treatment methods. J. Waste Manage. Assoc.. 63, 431–434 (2010) Available in Japanese
12. Shimizu, H, Nagata, K: Integrated life cycle assessment (LCA) approach for printing service by using environmental load point (ELP) method. Soc. Printing. Sci. Technol.. 47, 177–185 (2010)
13. Nagata, K, Ureshino, M, Matsuda, D, Ishikawa, M: Development of impact assessment in LCA: Characterization of category importance based on questionnaires, The Japan Society of Mechanical Engineering, 6th Environmental Engineering Symposium, Tokyo International Forum, Tokyo (1997)
14. Swiss Centre for Life Cycle Inventories: EcoInvent Database (2010) Accessed 10 June 2012
15. Wirawat, C, Gheewala, SH: Life cycle assessment of MSW-to-energy schemes in Thailand. J. Clean Prod.. 15, 1463–1468 (2007).
16. Williams, PT: Waste treatment and disposal, Wiley, West Sussex (2005)
17. The World Bank: Municipal solid waste incineration, World Bank Technical Guidance Report, Washington D.C. (1999)
18. Zhu, D, Asnani, PU, Zurbrugg, C, Anapolsky, S, Mani, S: Improving solid waste management in India, a source for policy makers and practitioners, Washington D.C., World Bank Institute (2009)
19. Hargreaves, JC, Adl, MS, Warman, PR: A review of the use of composted municipal solid waste in agriculture. Agric Ecosyst Environ. 123, 1–14 (2008).
20. Ali, G, Nitivattananon, V, Abbad, M, Sabir, M: Green waste to biogas: renewable energy possibilities for Thailand's green markets. 16, 5423–5429 (2012)
21. Ahsan, N: Solid waste management plan for Indian megacities. J. Environ. Prot.. 19, 90–95 (1999)
22. Tanoto, Y, Wijaya, ME: Economic and environmental emissions analysis in Indonesian electricity expansion planning: low-rank coal and geothermal energy utilization scenarios. Intern. J. Renew Energy Res.. 1, 61–66 (2011)
23. Gunamantha, M: Sarto: Life cycle assessment of municipal solid waste treatment to energy options: case study of Kartamantul region, Yogyakarta. Renew Energy. 41, 277–284 (2012)
24. Xianghua, D, Nie, Z, Yuan, B, Zuo, T: LCA case studies life cycle inventory for electricity generation in China. Int. J. LCA. 4, 217–224 (2007)
25. Chakraborty, N, Mukherjee, I, Santra, AK, Chowdhury, S, Chakraborty, S, Bhattacharya, S, Mitra, AP, Sharma, C: Measurement of CO_2, CO, SO_2, and NO emissions from coal-based thermal power plants in India. Atmos. Environ.. 42, 1073–1082 (2008).
26. Corporation, IEC: India Energy Handbook: Demand Driven, Supply Chained, PSI Media Inc, Delhi (2011)

27. US Energy Environment Administration: International energy statistics (2012) http://www.eia.gov/countries/data.cfm. webcite Accessed 10 June 2012
28. The Ministry of Environment: Indonesia: Emisi Gas Rumah Kaca Dalam Angka (GHG emission in figure), The Republic of Indonesia, Jakarta (2009)
29. United Nations: Statistics Division: Green house gases emissions
30. Zhang, Q, Streets, DG, Carmichael, GR, He, KB, Huo, H, Kannari, A, Klimont, Z, Park, IS, Reddy, S: Asian emissions in 2006 for the NASA INTEX-B mission. Atmospheric Chem. Phys.. 9, 5131–5153 (2006)
31. National Statistics Office, Indonesia: Mining production 1996–2010 http://www.bps.go.id/tab_sub/view.php?kat=3&tabel=1&daftar=1&id_subyek=10&notab=3 (2012). Accessed 6 June 2012
32. China Mining Association (CMA): Supply and demand
33. Index mundi: Mining Product Primary Production by Year
34. The Ministry of Environment, Indonesia: Indonesian Domestic Solid Waste Statistics, The Republic of Indonesia, Jakarta (2008)

PART II

ANAEROBIC DIGESTION

CHAPTER 3

Utilization of Household Food Waste for the Production of Ethanol at High Dry Material Content

LEONIDAS MATSAKAS, DIMITRIS KEKOS, MARIA LOIZIDOU, AND PAUL CHRISTAKOPOULOS

3.1 BACKGROUND

The environmental crisis and the shortage of fossil fuels have turned public attention to the utilization of other forms of energy, which are environmentally friendly and renewable, such as bio-ethanol [1,2]. During recent years research has focused on the so-called second-generation biofuels, where wastes or by-products are utilized as raw material, compared to the first-generation biofuels where sugars and starch were utilized. This way, increasing pubic concerns about utilizing food sources for the production of biofuels can be solved, as the utilization of either sugars or corn for the production of biofuels have contributed to the increase of their price

---

Utilization of Household Food Waste for the Production of Ethanol at High Dry Material Content.
© *Matsakas L, Kekos D, Loizidou M, and Christakopoulos P.* Biotechnology for Biofuels 7,4 (2014). *doi:10.1186/1754-6834-7-4. Licensed under a Creative Commons Attribution 2.0 Generic License, http://creativecommons.org/licenses/by/2.0/.*

worldwide, resulting in severe problems for the poorer countries. All these concerns led to a rapid increase in research to utilize low-cost by-products and wastes as raw material [3-6]. Lignocellulosic biomass represents great potential to be utilized as raw material due to the high amounts produced every year [7], and can be derived from woody or agricultural residues such as wheat straw, corn cobs, bagasse, rice straw, et cetera. The main challenge of utilizing lignocellulosic biomass is efficient sugar release, mainly from cellulose. In order to achieve this, an efficient pretreatment step followed by enzymatic hydrolysis have to be applied [8,9]. Generally, the pretreatment process contributes to increased costs of the whole process [10].

A different and alternative source of raw material for the production of biofuels could be the utilization of municipal organic wastes and especially household food wastes (HFW). Taking into account that the total quantity of HFW for the EU-27 during 2006 is estimated to be 37.7 Mt, which accounts for approximately 76 kg per capita and represents 42% of the total amount of food wastes generated in the EU [11], it is clear that they represent a challenge concerning their disposal, as well as an attractive raw material for the production of biofuels. Moreover, there is a trend of increasing the quantities of total food wastes produced (which are coming from both domestic, manufacture, food service/catering and retail/wholesale sectors), which, according to the European Commission (EC), will rise from 89.3 Mt in 2006 to 126.2 Mt by 2020 [11]. A common practice of HFW management is landfill disposal, which is causing severe environmental problems (such as greenhouse gas emissions) and shortage of disposal places [6,12,13]. Other practices are utilization as animal feed (which can raise hygiene issues) and soil conditioners-fertilizers (which can cause severe pollution to surface and underground water) [6,12,14]. Alternatively HFW can be used for the production of bio-based (green) chemicals and bio-energy (for example, biogas and ethanol) [15,16]. Until now, most of the research dedicated to HFW utilization was focused on biogas production.

Utilization of HFW as raw material represents a great challenge, as both the collection of generated HFW from multiple places [13] and post-collection treatment are difficult processes. Moisture and soluble sugars can make HFW an easy target for microorganisms, leading to their severe degradation. Another challenge is the heterogeneity that HFW present [13,14], which is highly affected by the source from which the wastes are derived. Nutritional habits and season of collection can also affect the composition of the HFW. Generally, fruits and vegetables represent a significant portion of the wastes [6,17]. Finally, one important issue to be solved is the proper education of the public in order to achieve low presence of contaminants (for example, plastics, metal et cetera) during source separation of HFW.

Concerning the utilization of HFW, there are some reports where different types of pretreatment, such as acid, alkali and thermal, have been used in order to increase cellulose digestibility [18-20]. Despite the fact that a pretreatment process can increase digestibility of cellulose, the soluble sugars can be degraded forming various inhibitors (such as furfural), especially if the pretreatment is performed at harsher conditions and in the presence of alkali.

The aim of this work was the utilization of source-separated HFW for the production of ethanol, at high dry material (DM) levels in order to achieve high ethanol production. Utilization of HFW at high DM levels results in a very viscous mash, where only solid-state cultivation can be applied, which presents many disadvantages including difficulties for process scaling-up and ethanol recovery [21]. In order to overcome this obstacle, an enzymatic liquefaction/saccharification process prior to fermentation, employing commercial cellulases solution (Celluclast®1.5 L and Novozym 188) was applied. During this process the viscosity of the high solid-content substrate was rapidly reduced, enabling submerged fermentation. No pretreatment prior to enzymatic saccharification was applied, in order to minimize the soluble sugar degradation. Finally, in order to maximize the ethanol yield a subsequent treatment and fermentation of the remaining solids (residue) was applied. During recent years, a new gravimetric mixing system has been successfully applied for the liquefaction of pretreated lignocellulosic feedstock at high DM content and was used in the present study [22,23].

3.2 RESULTS AND DISCUSSION

3.2.1 FERMENTATION OF SACCHARIFIED HFW

Table 1 shows the HFW composition obtained. HFW has potential to be utilized as raw material, as cellulose content is quite high and soluble sugars, such as glucose, fructose and sucrose are present and can be readily converted to ethanol. According to the literature, most researchers are analyzing food wastes by measuring the chemical oxygen demand (COD), biological oxygen demand (BOD), volatile solids (VS) et cetera [18-20,24-26], especially when they are utilized for biogas production. Though these values can provide important information about the raw material, for the ethanol production processes it is more important to know the proportion of soluble and insoluble sugars, as well as the type of insoluble polysaccharides, in order to apply the most appropriate enzymatic hydrolysis treatment.

TABLE 1: Composition of HFW

Fraction	% w/w
Soluble	33.81±0.42
Glucose	4.39±0.20
Fructose	3.47±0.12
Sucrose	4.38±0.10
Total reducing sugars	12.54±0.93
Protein	0.54±0.01
Fats	11.91±0.68
Crude protein	10.51±0.37
Pectin	3.92±0.33
Cellulose	18.30±0.19
Hemicellulose	7.55±0.39
Klason lignin	2.16±0.25
Ash	11.03±0.42

Initial moisture content was 1.03±0.20% w/w. The values are given as mean±SD.

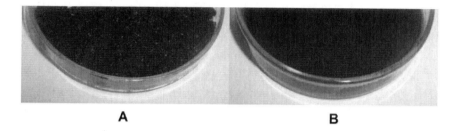

FIGURE 1: Effect of enzymatic liquefaction/saccharification on household food wastes (HFW) at 45% dry material (DM). (A) Prior to liquefaction/saccharification; (B) after 8 h of liquefaction/saccharification. Liquefaction of HFW was conducted for 8 h at an initial DM content of 45% w/v at 50°C. The enzyme load applied was 10 unit/g DM of a mixture of Celluclast® 1.5 L and Novozym 188 at a ratio of 5:1 v/v.

Composition of food wastes can present a wide variety. Zhang and Richard [27] utilized a food-waste sample from a composting site of a University with a composition of 23.3% w/w total reducing sugars, 34.8% w/w starch and 1.6% w/w fibers and employed mainly amylases for its saccharification. Moon et al.[6] mentioned a high starch (30.1% w/w) and fiber content (14.9% w/w) with total reducing sugars of 17.6% w/w making it necessary to utilize both amylases and cellulases to treat it, whereas a high starch content (63.9% w/w) combined with low cellulose amounts was reported by Yan et al.[12] for the HFW sample that was used in their experiments.

The liquefaction/saccharification process was performed for 8 h at an initial DM content of 45% w/v followed by fermentation at two different initial DM contents (35% and 45% w/v). As can been evidenced from Figure 1, HFW was fully liquified after 8 hours of enzymatic treatment. This fact is also supported by the difference in the viscosity measured at an angular velocity of 10 rad/s before and after the enzymatic treatment, which decreased from 2790 Pa·s to 67.5 Pa·s, respectively. Maximum ethanol production in both DM contents was observed after 15 h of fermentation (Figure 2) and found to be 34.85 g/L (35% w/v DM) and 42.78 g/L (45% w/v DM) with a volumetric productivity of 2.32 g/L·h (35% w/v DM) and 2.85 g/L·h (45% w/v DM) (Table 2). Cellulose hydrolysis at the end of the fermentation reached 50.27% of the initial cellulose content in raw material. Considering this, the obtained yields (Yp/s) were 0.443 g/g and 0.423 g/g at 35% and 45% w/v DM respectively. The highest ethanol yield obtained when the fermentation was performed at 35% w/v DM could be attributed to the better mixing conditions.

TABLE 2: Results of ethanol production during cultivation of *S. cerevisiae* on HFW

Initial dry material	Separate liquefaction step	Ethanol production	Ethanol productivity	% of the maximum theoretical yield[a]	% of the maximum theoretical yield[b]
(% w/v)		(g/L)	(g/L·h)		
35	+	34.85±0.55	2.32±0.04	59.82±0.94	159.21±2.51
45	+	42.78±0.83	2.85±0.06	57.12±1.10	151.91±2.93
35	-	24.75±2.20	0.52±0.05	42.48±3.78	113.07±10.05
45	-	39.15±0.75	0.82±0.02	52.28±1.00	139.07±2.66

% Maximum theoretical ethanol yield, calculations were; [a]based on the maximum ethanol that could be produced from the soluble and the cellulosic sugars; [b]based on the maximum ethanol production from the soluble sugars only. The values are given as mean±SD.

With no use of the separate liquefaction step the ethanol production reduced by 28.98% and 8.49% when fermentation performed at 35% and 45% w/v initial DM, respectively. The use of the liquefaction/saccharification step was also associated with a significant increase in ethanol volumetric productivities (Table 2), mainly due to the partial cellulose hydrolysis which enabled reduction of viscosity and better mixing conditions of the fermenting raw material [28]. Same enhancement in ethanol production efficiency was also demonstrated by Kim et al.[25], who observed an increase in ethanol yield from 0.31 g/g to 0.43 g/g total solid when applying a Separate Hydrolysis and Fermentation (SHF) process instead of Simultaneous Saccharification and Fermantation (SSF) process on cafeteria food waste. Manzanares et al.[29] also found that with increasing initial DM content, a separate saccharification step improves fermentation of liquid hot water-pretreated olive-pruning biomass. Finally, Hoyer et al.[30]

reported that with increasing DM content, even 4 h of saccharification could significantly improve fermentation of softwoods.

Ethanol production efficiency during this work was higher than that compared to Moon et al.[6] who performed a 3-h liquefaction process of food waste using both carbohydrases and amyloglucosidases where the ethanol production reached 29.1 g/L (Table 3). Walker et al.[31] utilized food wastes from starch-containing food and after saccharification with amylases the overall ethanol production was 8 g/L. Uncu and Cekmecelioglu [14] achieved 32.2 g/L ethanol production after 59 h of fermentation using food wastes treated for 6 h with amylases. Jeong et al.[32] reached 40.59 g/L ethanol production after 24 h of fermentation on food wastes hydrolyzed for 8 h with enzymes, using the fermentative microorganism *Saccharomyces coreanus*. When *Pichia stipitis* was added as a co-fermenting microorganism, ethanol production increased up to 48.63 g/L but the obtained productivities were lower than those of the present work. Cekmecelioglu and & Uncu [33] reported an ethanol production of 23.3 g/L after 48 h of cultivation on kitchen wastes saccharified for 6 h. Yan et al.[12] reported an ethanol production of 81.5 g/L from a saccharified high starch containing raw material (starch content was 63.9% w/w) using a high glucoamylase load (142.2 unit/g). Finally, Kim et al.[34] achieved 57.5 g/L ethanol production after 14 h of fermentation using starchy food waste saccharified for 4 h with amylases.

3.2.2 PRETREATMENT AND FERMENTATION OF THE SOLID RESIDUE

At the end of the fermentation there is a remaining solid fraction that was not converted to ethanol. This solid fraction contains unhydrolyzed cellulose which practically is lost from the ethanol production process. In order to increase the overall biofuel yield of the raw material's mass, these solids could be further utilized. Some research has proposed the utilization of the remaining solids after fermentation and ethanol distillation for the production of biogas on kitchen waste [35], oat straw [36], wheat straw [37,38] and corn stover [39].

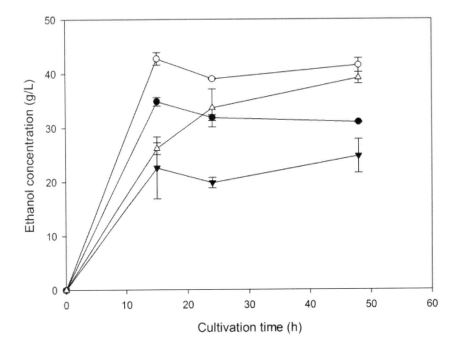

FIGURE 2: Production of ethanol from liquefied household food waste (HFW). Time course of ethanol production from HFW at 35% (solid circles) and 45% (open circles) initial dry material (DM) content with liquefaction/saccharification and at 35% (solid triangles) and 45% (open triangles) without liquefaction/saccharification.

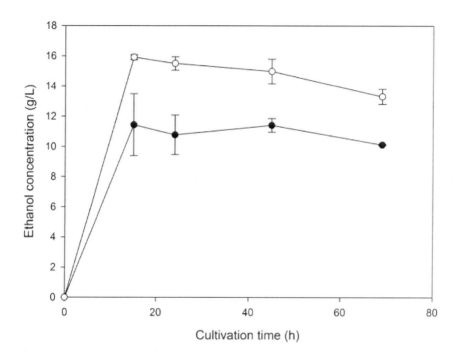

FIGURE 3: Production of ethanol from liquefied residue. Time course of ethanol production from hydrothermally pretreated residue at 35% (solid circles) and 45% (open circles) initial dry material content.

TABLE 3: Production of ethanol from food wastes from different sources

Source of food waste	Carbohydrate content (% w/w)		Ethanol yield parameters		Reference
	Soluble[a]	Fiber	Concentration (g/L)	Productivity (g/L·h)	
Cafeteria	n.a.	n.a.	n.a.	n.a.	[25]
Cafeteria	47.7	14.9	29.1	1.94	[6]
Dining center	n.a.	n.a.	8	n.a.	[31]
Cafeteria and houses	69		32.2	0.55	[14]
Cafeteria	n.a.	n.a.	48.63	2.03	[32]
Food courts	57.6		23.3	0.49	[33]
Dining room	63.9	1.98	81.5	1.36	[12]
Cafeteria	n.a.	n.a.	57.5	4.11	[34]
Houses	12.24	18.30	42.78	2.85	This work
Houses	12.24	18.30	34.85	2.32	This work

[a]Both soluble sugars and starch; n.a., not available.

As it is described in the Methods section, after the end of the fermentation at 45% initial DM the solids were removed from the fermentation broth. This solid fraction contains the unhydrolyzed cellulose fraction, which could be further utilized for the production of ethanol in order to increase the overall production yield. The high degree of recalcitrance of this fraction [40] makes a pretreatment process prior to liquefaction/saccharifiacation necessary. During this study, hydrothermal pretreatment with the presence of acetic acid as a catalyst was applied [41,42]. After the pretreatment, solids were separated from the liquid fraction and washed with distilled water in order to remove the catalyst and other inhibitors formed during pretreatment. Inhibitor removal is necessary in order to decrease the stress to the fermenting microorganism, allowing higher fermentation rates and ethanol production efficiency [43-47].

As has been previously discussed cellulose hydrolysis reached 50.27% of the initial presented cellulose in HFW whereas the cellulose content of residue was 14.75% w/w. During the hydrothermal pretreatment process 42.73% of the initial mass of the residue was solubilized. Cellulose content of the pretreated residue was 16.31%, indicating a 36.68% of cellulose solubilization during the pretreatment.

During fermentation of the pretreated residue at 35% and 45% w/v, maximum ethanol concentrations of 11.44 g/L and 15.92 g/L, respectively, were observed after 15 h (Figure 3). Moreover, increasing substrate concentrations of the fermented residue was associated with an increase in ethanol productivity (Table 4). From the initial cellulosic fraction of the pretreated residue, 42.67% was hydrolyzed. The obtained yields (Yp/s) reached 0.423 g/g and 0.458 g/g at 35% and 45% DM respectively, which were almost identical with the yields obtained during fermentation of HFW.

TABLE 4: Results of ethanol production during cultivation of S. cerevisiae on pretreated residue

Initial DM (% w/v)	Ethanol (g/L)	Productivity (g/L·h)	% of maximum theoretical
35	11.44 ± 1.45	0.76 ± 0.10	35.34 ± 4.49
45	15.92 ± 0.12	1.06 ± 0.01	38.23 ± 0.28

The values are given as mean ± SD.

3.2.3 OVERALL ETHANOL YIELDS

Figure 4 presents the overall obtained ethanol yield after the fermentation of 1 kg of raw material using the two-stage sequential fermentation procedure (concerning the fermentations at 45% DM). In the first stage, 95.07 g of ethanol and 617.2 g of residue were obtained. The remaining residue was hydrothermally pretreated and the solid fraction was 353.5 g, whereas the other solids were dissolved to the liquid fraction. In the second stage after fermentation of the pretreated residue 12.51 g of ethanol could be obtained, thus the ethanol production yield could be increased from 95.07 g/kg DM to 107.58 g/kg DM, corresponding to an increase of 13.16%. The ethanol yield comparing to the maximum theoretical increased from 57.12% to 63.64%.

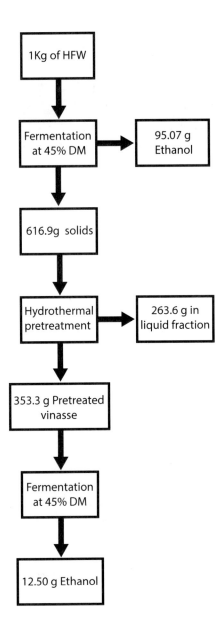

FIGURE 4: Overall ethanol production yield. Ethanol production yield after fermentation of liquefied household food wastes (HFW) at initial dry material (DM) content of 45% and subsequent fermentation of the residue. Prior to fermentation, residue was hydrothermally pretreated at 200°C for 10 minutes and liquefied.

3.2.4 CONCLUSIONS

In the current work the potential of utilizing source-separated HFW for the production of ethanol at high DM content was demonstrated. A liquefaction process prior to fermentation increased both ethanol production and ethanol volumetric productivity. Finally the subsequent fermentation of residue increased the overall ethanol production yield.

3.3 METHODS

3.3.1 RAW MATERIAL

HFW utilized as raw material during this work were source-separated from houses in Papagos-Cholargos Municipality, Athens, Greece. The wastes were dried in situ in a prototype house dryer designed and developed by the Unit of Environmental Science and Technology (UEST), School of Chemical Engineering, NTUA [48]. Dried HFW were milled with a small laboratory mill at an average particle size less than 3 mm. The composition of dried HFW is presented in Table 1.

3.3.2 REAGENTS AND ENZYME SOLUTIONS

All chemicals were of analytical grade. During this work a mixture of the commercial enzyme solutions from Novozymes A/S (Bagsværd, Denmark) Celluclast® 1.5 L (cellulases) and Novozym 188 (β-glucosidase) at a ratio of 5:1 v/v has been applied for the liquefaction/saccharification process. The activity of the mixture was measured according to the standard filter paper assay [49] and found to be 83 FPU/mL.

3.3.3 ANALYTICAL METHODS

Total reducing sugars were measured by the dinitro-3,5-salicilic acid (DNS) method [50]. Monomeric sugars and ethanol were analyzed by

HPLC (Shimadzu LC-20 AD, Kyoto, Japan) equipped with a refractive index detector (Shimadzu RID 10A, Kyoto, Japan). Monomeric sugar were analyzed utilizing an Aminex HPX-87P (300 × 7.8 mm, particle size 9 μm, Bio-Rad, Hercules, California) chromatography column, operating at 70°C with HPLC-water as a mobile phase at a flow rate of 0.6 mL/minute. Ethanol was determined by an Aminex HPX-87H (300 × 7.8 mm, particle size 9 μm, Bio-Rad, Hercules, California) chromatography column at 40°C, with a mobile phase of 5 mM sulfuric acid (H2SO4) at a flow rate of 0.6 mL/minute.

Soluble fraction was analyzed according to the official method of the National Renewable Energy Laboratory (NREL) [51]. The liquid fraction was further analyzed for sugars and proteins [52]. Moisture was analyzed according to Sluiter et al.[53], whereas crude fat, ash, protein and total starch content determination were conducted according to standard Association of Official Agricultural Chemists (AOAC) methods [54]. Pectin was determined according to Phatak et al. [55]. Finally, the cellulose, hemicellulose and (acid-insoluble) lignin content was determined according to Sluiter et al.[53]. It is worth mentioning that the HFW utilized during this work had no starch content. Analysis was carried out in triplicate. Apparent viscosity of the HFW before and after enzymatic treatment was determined by an Anton Paar Physica MCR rheometer apparatus (Anton Paar GmbH, Ashland, USA) as previously described [56].

3.3.4 HYDROTHERMAL PRETREATMENT OF RESIDUE

The remaining solids after fermentation of HFW were hydrothermally pretreated by microwave digestion equipment at 200°C for 10 minutes as previously described [41]. After the pretreatment, the solids were removed from the liquid fraction through vacuum filtration and washed in order to remove inhibitors formed during pretreatment. Finally, solids were dried at 60°C until constant weight reached.

3.3.5 ENZYMATIC LIQUEFACTION AND SACCHARIFICATION OF HFW

Enzymatic liquefaction/saccharification of untreated HFW and hydrothermally pretreated residue was conducted in a liquefaction reactor which was designed and manufactured in-house. More specifically, the reactor consists of two vertical cylindrical chambers which are 6 cm wide and 25 cm diameter, with a rotating shaft driven by a 0.37-kW motor for the mixing of the material. The mixing shaft was programmed to shift from clockwise to anti-clockwise rotation every minute in order to achieve better mixing. Finally, the temperature was controlled by an external oil jacket.

The liquefaction/saccharification process was performed at initial DM concentration of 45% w/v for 8 h. The pH was adjusted to 5.0 by using 50 mM citrate-phosphate buffer and the enzyme load applied was 10 FPU/g DM. Finally, the temperature of the liquefaction/saccharification was set at 50°C. At the end of the liquefaction/saccharification process the whole slurry (also containing unhydrolyzed solids) was utilized for the fermentation experiments.

3.3.6 ETHANOL FERMENTATION

Fermentations at 35% and 45% (w/v) DM of non-sterilized liquefied HFW or pretreated residue were performed in 100-mL Erlenmeyer flasks in an orbital shaker at 30°C with an agitation of 100 rpm. The fermenting microorganism was dry baker's yeast (Yiotis, Athens, Greece), which was added at a concentration corresponding to 15 mg/g of initial DM. To evaluate the importance of the separate liquefaction/saccharification step, untreated HFW were fermented under the same conditions. Samples were taken at certain time intervals, centrifuged and analyzed for ethanol. All trials were carried out in duplicate.

When the HFW fermentation process was completed, the broth was filtrated under vacuum in order to remove the solids which were further washed with distilled water. The solids (residue) were dried at 60°C until constant weight reached and were further utilized for ethanol production after being hydrothermally pretreated (as previously described).

REFERENCES

1. Li X, Kim TH, Nghiem NP: Bioethanol production from corn stover using aqueous ammonia pretreatment and two-phase simultaneous saccharification and fermentation (TPSSF). Bioresour Technol 2010, 101:5910-5916.
2. Sarris D, Giannakis M, Philippoussis A, Komaitis M, Koutinas AA, Papanikolaou S: Conversions of olive mill wastewater-based media by Saccharomyces cerevisiae through sterile and non-sterile bioprocesses. J Chem Technol Biotechnol 2013, 88:958-969.
3. Matsakas L, Christakopoulos P: Optimization of ethanol production from high dry matter liquefied dry sweet sorghum stalks. Biomass Bioenerg 2013, 51:91-98.
4. Yan S, Chen X, Wu J, Wang P: Ethanol production from concentrated food waste hydrolysates with yeast cells immobilized on corn stalk. Appl Microbiol Biotechnol 2012, 94:829-838.
5. Sims REH, Mabee W, Saddler JN, Taylor M: An overview of second generation biofuel technologies. Bioresour Technol 2010, 101:1570-1580.
6. Moon HC, Song IS, Kim JC, Shirai Y, Lee DH, Kim JK, Chung SO, Kim DH, Oh KK, Cho YS: Enzymatic hydrolysis of food waste and ethanol fermentation. Int J Energ Res 2009, 33:164-172.
7. Zhang M, Wang F, Su R, Qi W, He Z: Ethanol production from high dry matter corncob using fed-batch simultaneous saccharification and fermentation after combined pretreatment. Bioresour Technol 2010, 101:4959-4964.
8. Silva VN, Arruda P, Felipe MA, Gonçalves A, Rocha GM: Fermentation of cellulosic hydrolysates obtained by enzymatic saccharification of sugarcane bagasse pretreated by hydrothermal processing. J Ind Microbiol Biotechnol 2011, 38:809-817.
9. Díaz MJ, Cara C, Ruiz E, Romero I, Moya M, Castro E: Hydrothermal pre-treatment of rapeseed straw. Bioresour Technol 2010, 101:2428-2435.
10. Pérez JA, Ballesteros I, Ballesteros M, Sáez F, Negro MJ, Manzanares P: Optimizing liquid hot water pretreatment conditions to enhance sugar recovery from wheat straw for fuel-ethanol production. Fuel 2008, 87:3640-3647.
11. European Communities: EC preparatory study on food waste in the EU27. [http://ec.europa.eu/environment/eussd/pdf/bio_foodwaste_report.pdf]
12. Yan S, Li J, Chen X, Wu J, Wang P, Ye J, Yao J: Enzymatical hydrolysis of food waste and ethanol production from the hydrolysate. Renew Energ 2011, 36:1259-1265.
13. Lin CSK, Pfaltzgraff LA, Herrero-Davila L, Mubofu EB, Abderrahim S, Clark JH, Koutinas AA, Kopsahelis N, Stamatelatou K, Dickson F, Thankappan S, Mohamed Z, Brocklesby R, Luque R: Food waste as a valuable resource for the production

of chemicals, materials and fuels. Current situation and global perspective. Energ Environ Sci 2013, 6:426-464.
14. Uncu ON, Cekmecelioglu D: Cost-effective approach to ethanol production and optimization by response surface methodology. Waste Manage 2011, 31:636-643.
15. Luque R, Clark J: Valorisation of food residues: waste to wealth using green chemical technologies. Sustain Chem Process 2013, 1:10.
16. Arancon RAD, Lin CSK, Chan KM, Kwan TH, Luque R: Advances on waste valorization: new horizons for a more sustainable society. Energ Sci Eng 2013, 1:53-71.
17. Jensen JW, Felby C, Jørgensen H, Rønsch GØ, Nørholm ND: Enzymatic processing of municipal solid waste. Waste Manage 2010, 30:2497-2503.
18. Ma J, Duong TH, Smits M, Verstraete W, Carballa M: Enhanced biomethanation of kitchen waste by different pre-treatments. Bioresour Technol 2011, 102:592-599.
19. Singhal S, Bansal SK, Singh R: Evaluation of biogas production from solid waste using pretreatment method in anaerobic condition. Int J Emerg Sci 2012, 2:405-414.
20. Vavouraki AI, Angelis EM, Kornaros M: Optimization of thermo-chemical hydrolysis of kitchen wastes. Waste Manage 2013, 33:740-745.
21. Singhania RR, Patel AK, Soccol CR, Pandey A: Recent advances in solid-state fermentation. Biochem Eng J 2009, 44:13-18.
22. Jørgensen H, Vibe-Pedersen J, Larsen J, Felby C: Liquefaction of lignocellulose at high-solids concentrations. Biotechnol Bioeng 2007, 96:862-870.
23. Larsen J, Østergaard Petersen M, Thirup L, Wen Li H, Krogh Iversen F: The IBUS process – lignocellulosic bioethanol close to a commercial reality. Chem Eng Technol 2008, 31:765-772.
24. Bernstad A, Malmquist L, Truedsson C, la Cour Jansen J: Need for improvements in physical pretreatment of source-separated household food waste. Waste Manage 2013, 33:746-754.
25. Kim JH, Lee JC, Pak D: Feasibility of producing ethanol from food waste. Waste Manage 2011, 31:2121-2125.
26. Le Man H, Behera SK, Park HS: Optimization of operational parameters for ethanol production from Korean food waste leachate. Int J Environ Sci Tech 2010, 7:157-164.
27. Zhang X, Richard T: Dual enzymatic saccharification of food waste for ethanol fermentation. Proceedings of international conference on electrical and control engineering: 16–18 September 2011; Yichang ISBN 978-1-4244-8162-0
28. Szijarto N, Horan E, Zhang J, Puranen T, Siika-aho M, Viikari L: Thermostable endoglucanases in the liquefaction of hydrothermally pretreated wheat straw. Biotechnol Biofuels 2011, 4:2.
29. Manzanares P, Negro MJ, Oliva JM, Saéz F, Ballesteros I, Ballesteros M, Cara C, Castro E, Ruiz E: Different process configurations for bioethanol production from pretreated olive pruning biomass. J Chem Technol Biotechnol 2011, 86:881-887.
30. Hoyer K, Galbe M, Zacchi G: Production of fuel ethanol from softwood by simultaneous saccharification and fermentation at high dry matter content. J Chem Technol Biotechnol 2009, 84:570-577.
31. Walker K, Vadlani P, Madl R, Ugorowski P, Hohn KL: Ethanol fermentation from food processing waste. Environ Prog Sustain Energ 2012, 32:1280-1283.

32. Jeong S-M, Kim Y-J, Lee D-H: Ethanol production by co-fermentation of hexose and pentose from food wastes using Saccharomyces coreanus and Pichia stipitis. Korean J Chem Eng 2012, 29:1038-1043.
33. Cekmecelioglu D, Uncu ON: Kinetic modeling of enzymatic hydrolysis of pretreated kitchen wastes for enhancing bioethanol production. Waste Manage 2013, 33:735-739.
34. Kim JK, Oh BR, Shin H-J, Eom C-Y, Kim SW: Statistical optimization of enzymatic saccharification and ethanol fermentation using food waste. Process Biochem 2008, 43:1308-1312.
35. Tang Y-Q, Koike Y, Liu K, An M-Z, Morimura S, Wu X-L, Kida K: Ethanol production from kitchen waste using the flocculating yeast Saccharomyces cerevisiae strain KF-7. Biomass Bioenerg 2008, 32:1037-1045.
36. Dererie DY, Trobro S, Momeni MH, Hansson H, Blomqvist J, Passoth V, Schnürer A, Sandgren M, Ståhlberg J: Improved bio-energy yields via sequential ethanol fermentation and biogas digestion of steam exploded oat straw. Bioresour Technol 2011, 102:4449-4455.
37. Bauer A, Bösch P, Friedl A, Amon T: Analysis of methane potentials of steam-exploded wheat straw and estimation of energy yields of combined ethanol and methane production. J Biotechnol 2009, 142:50-55.
38. Kaparaju P, Serrano M, Thomsen AB, Kongjan P, Angelidaki I: Bioethanol, biohydrogen and biogas production from wheat straw in a biorefinery concept. Bioresour Technol 2009, 100:2562-2568.
39. Bondesson PM, Galbe G, Zacchi G: Ethanol and biogas production after steam pretreatment of corn stover with or without the addition of sulphuric acid. Biotechnol Biofuels 2013, 6:11.
40. Xiros C, Katapodis P, Christakopoulos P: Evaluation of Fusarium oxysporum cellulolytic system for an efficient hydrolysis of hydrothermally treated wheat straw. Bioresour Technol 2009, 100:5362-5365.
41. Matsakas L, Christakopoulos P: Fermentation of liquefacted hydrothermally pretreated sweet sorghum bagasse to ethanol at high-solids content. Bioresour Technol 2013, 127:202-208.
42. Petrik S, Kádár Z, Márová I: Utilization of hydrothermally pretreated wheat straw for production of bioethanol and carotene-enriched biomass. Bioresour Technol 2013, 133:370-377.
43. da Cunha-Pereira F, Hickert LR, Sehnem NT, de Souza-Cruz PB, Rosa CA, Ayub MAZ: Conversion of sugars present in rice hull hydrolysates into ethanol by Spathaspora arborariae, Saccharomyces cerevisiae, and their co-fermentations. Bioresour Technol 2011, 102:4218-4225.
44. Alvira P, Moreno AD, Ibarra D, Sáez F, Ballesteros M: Improving the fermentation performance of Saccharomyces cerevisiae by laccase during ethanol production from steam-exploded wheat straw at high-substrate loadings. Biotechnol Prog 2013, 29:74-82.
45. Geddes CC, Peterson JJ, Roslander C, Zacchi G, Mullinnix MT, Shanmugam KT, Ingram LO: Optimizing the saccharification of sugar cane bagasse using dilute phosphoric acid followed by fungal cellulases. Bioresour Technol 2010, 101:1851-1857.

46. Palmqvist E, Hahn-Hägerdal B: Fermentation of lignocellulosic hydrolysates. II: inhibitors and mechanisms of inhibition. Bioresour Technol 2000, 74:25-33.
47. Klinke HB, Thomsen AB, Ahring BK: Inhibition of ethanol-producing yeast and bacteria by degradation products produced during pre-treatment of biomass. Appl Microbiol Biotechnol 2004, 66:10-26.
48. LIFE 08/ENV/GR/000566 [http://www.uest.gr/drywaste/site/index.htm]
49. Ghose TK: Measurement of cellulase activities. Pure Appl Chem 1987, 59:257-268.
50. Miller GL: Use of dinitrosalicylic acid reagent for determination of reducing sugar. Anal Chem 1959, 31:426-428.
51. Sluiter A, Ruiz R, Scarlata C, Sluiter J, Templeton D: Determination of Extractives in Biomass. Technical report NREL/TP-510-42619, Laboratory analytical protocol. Golden CO: National Renewable Energy Laboratory; 2008.
52. Bradford MM: A rapid and sensitive method for the quantification of microgram quantities of protein utilizing the principle of protein-dye binding. Anal Biochem 1976, 72:248-254.
53. Sluiter A, Hames B, Ruiz R, Scarlata C, Sluiter J, Templeton D, Crocker D: Determination of structural carbohydrates and lignin biomass. Technical report NREL/TP-510-42618, Laboratory analytical protocol. Golden CO: National Renewable Energy Laboratory; 2012.
54. William H: Official methods of analysis of the association of official analytical chemists. Washigton DC: AOAC Inc; 1970.
55. Phatak L, Chang KC, Brown G: Isolation and characterization of pectin in sugarbeet pulp. J Food Sci 1988, 53:830-833.
56. Karnaouri AC, Topakas E, Christakopoulos P: Cloning, expression, and characterization of a thermostable GH7 endoglucanase from Myceliophthora thermophila capable of high-consistency enzymatic liquefaction. Appl Microbiol Biotechnol 2013. in press

CHAPTER 4

Production of Fungal Glucoamylase for Glucose Production from Food Waste

WAN CHI LAM, DANIEL PLEISSNER, AND CAROL SZE KI LIN

4.1 INTRODUCTION

Food waste is a serious global problem, especially in many developed countries. In Hong Kong, over 3500 tons of food wastes are generated every day [1]. Currently, landfilling and incineration are the major practices for managing these wastes in many countries. These practices, however, may cause severe environmental pollutions and adds burden to the economy. Due to its high contents of carbohydrates and proteins, food wastes may serve as feedstock in biorefineries for production of fungal enzymes, e.g., glucoamylase (GA) and offers an innovative approach to waste management.

GA is a family of amylolytic enzymes that catalyze the cleavage of α-(1,4) glycosidic bonds in starch and release glucose as end product [2,3]. Glucose is the principle carbon source in many biotechnological processes and of great importance for fermentative chemicals and fuel production

---

Production of Fungal Glucoamylase for Glucose Production from Food Waste. © Lam WC, Pleissner D, and Lin CSK. Biomolecules 3,3 (2013). doi:10.3390/biom3030651. Licensed under a Creative Commons Attribution 3.0 Unported License, http://creativecommons.org/licenses/by/3.0/.

such as succinic acid, bio-plastic and ethanol. Starch is usually the major component of mixed food waste from restaurants [4,5,6], application of GA for food waste hydrolysis to recover glucose from starch, therefore, may not only offer a solution for managing food waste but also help to save precious resources.

Aspergillus awamori is a known secretor of GA with beneficial properties for industrial bioprocesses such as high productivity and enzyme activity at high temperatures [3,7]. Therefore, *Aspergillus awamori* is employed in this study for GA production through solid state fermentation (SSF). SSF is a fermentation process conducted in the absence of free water, thus it is desirable for industrial enzymes production since the enzymes produced at the end are not diluted by the amount water added in comparison to submerge fermentation. Consequently, the enzymes produced are at a much higher concentration. Pastry waste collected from local Starbucks contains 44.6% of starch and 7% of protein [8]. Starch has been shown as an inducer for GA synthesis by some *Aspergillus* producers [9,10,11]; thus, the significant amount of starch in pastry waste could be desirable for GA production. The smaller amount of protein, 7%, on the other hand could provide a source of nitrogen to promote the fungal growth and facilitate GA production. Therefore, pastry waste was selected in this study for GA production. The crude GA extract obtained without further purification was characterized in terms of optimal pH and reaction temperature, as well as thermo-stability. Additionally, application of the crude GA extract was studied for hydrolysis of mixed food waste collected from a local restaurant to produce high glucose solution.

4.2 RESULTS AND DISCUSSION

4.2.1 GLUCOAMYLASE PRODUCTION FROM PASTRY WASTE

To demonstrate the feasibility of pastry waste as the sole substrate for GA production, SSF was conducted with *Aspergillus awamori* without addition of any other nitrogen or carbon sources. GA production from pastry waste over time is shown in Figure 1. Production of GA reached maximal activity at Day 10. Approximately, 253.7 ± 20.4 U of GA was produced from one gram of pastry waste on dry basis (d.b.); the GA activity of the crude GA extract was 76.1 ± 6.1 U/mL. The result demonstrates that pastry waste can be used solely for GA production.

GA production with different wastes as substrates have been studied and the results are summarized in Table 1. In some studies, nitrogen in form of ammonium, urea and yeast extract was supplemented to substrates to facilitate fungal growth and GA production [4,12,13,14]. In contrast, nitrogen supplement was not involved in this study, but the GA activity of enzyme extract appears higher than in earlier studies with nitrogen supplement most likely due to a good balance of carbon (C) to nitrogen (N) and phosphorus (P) ratios. The data suggests that pastry waste is a promising substrate for GA production.

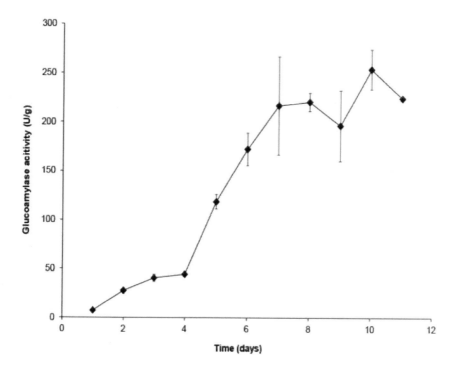

FIGURE 1: Glucoamylase (GA) production from pastry waste with *Aspergillus awamori* during solid state fermentation(SSF) for 11 days at 30°C. Experiments were duplicated. The mean values are plotted and the standard errors are reported.

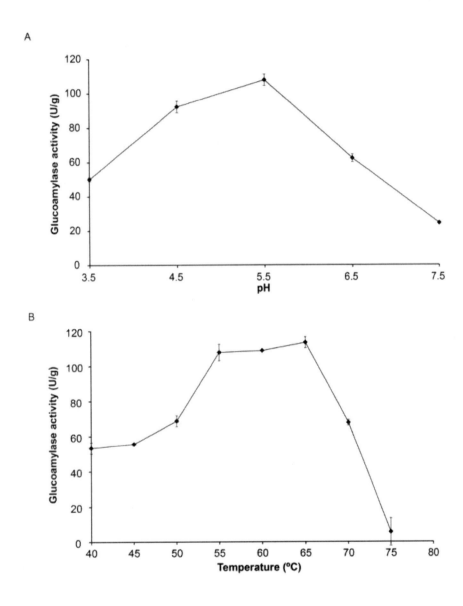

FIGURE 2: Effect of (A) pH at 55°C and (B) temperature at pH 5.5 on crude GA extract activity. Experiments were duplicated. The mean values are plotted and the standard errors are reported.

TABLE 1: GA production and yields from different studied substrates with or without nitrogen supplement through solid state fermentation.

Substrate	Crude GA concentration (U/mL)	Yield (U/g)	Fungus	Nitrogen supplement	References
Rice powder	N/A	71.3 ± 2.34 a	Aspergillus niger	+	[12]
Wheat bran	N/A	110 ± 1.32 a	Aspergillus niger	+	[12]
Mixed food waste	137	N/A	Aspergillus niger	+	[13]
Cowpea waste	970	N/A	Aspergillus oryzae	-	[14]
Wheat bran	4.4	48	Aspergillus awamori	-	[15]
Wheat pieces	3.32	81.3	Aspergillus awamori	-	[16]
Waste bread	3.94	78.4	Aspergillus awamori	-	[16]
Waste bread	N/A	114	Aspergillus awamori	-	[17]
Pastry waste	76.1 ± 6.1[a]	253.7 ± 20.4[a]	Aspergillus awamori	-	This study

[a] *Values indicate means ± standard errors*

4.2.2 CHARACTERIZATION OF OPTIMAL REACTION TEMPERATURE AND PH OF THE CRUDE GLUCOAMYLASE EXTRACT

Optimal pH and reaction temperature of the crude GA extract were determined. The results are shown in Figure 2. In order to determine the optimal reaction pH of the crude GA extract, assays were carried out at various pHs from 3.5 to 7.5 and the results are indicated in Figure 2A. The maximal enzyme activity was obtained at pH 5.5, indicating it was the optimal pH for starch hydrolysis. Fungal GAs from *Aspergillus* strains are usually active at acidic pH, their enzyme activities vary from pH 3.5 to 7 depending on the strains and amino acid sequences (isoform) [3].

Similarly, GA assay was conducted at different temperatures as indicated in Figure 2B from 40–75 °C with 5 °C increment at pH 5.5 in order to determine the optimal reaction temperature for the crude GA extract and assuming that the pH factor is independent of the temperature factor. A significant increase in enzyme activity was observed as temperature increases from 40 to 55 °C. Maximal GA activity was observed from 55 to 65 °C suggesting the range for optimal reaction temperature of the crude GA extract and in accordance with optimal reaction temperatures in the range of 50 to 60 °C usually found for GAs from *Aspergillus* [3]. Further increase in reaction temperature greatly reduced the enzyme activity most likely due to enzyme denaturation.

4.2.3 THERMO-STABILITY OF THE CRUDE GLUCOAMYLASE EXTRACT AT OPTIMAL REACTION TEMPERATURES

Since high GA activity was observed at 55, 60 and 65 °C, thermo-stability of the crude GA extract at these temperatures was further investigated in order to determine the optimal digestion temperature for the subsequent food waste hydrolysis experiment. The residual enzyme activity after heated at 55, 60 and 65 °C for over 90 min is shown in Figure 3. The rate constant (k_d, minutes^{-1}) for the first-order thermal deactivation was determined from the slope of the deactivation time course as shown in Figure 3 using Equation (1) [18], where E_t is the residual GA activity after heat treatment for time t. E_0 is the initial enzyme activity before heat treatment. The half-life of thermal deactivation ($t_{1/2}$) was determined according to Equation (2) [19]. The thermal deactivation of the crude GA extract exhibited a linear relationship showing that it followed first-order kinetics as reported [20]. The thermal deactivation rate constant k_d (minutes^{-1}) of the crude GA extract at 55 °C was found approximately 10 times slower than the rates at 60 and 65 °C, suggesting the crude GA is more thermo-stable at 55 °C. The kd of the crude GA extract at 55 °C was 2.20×10^{-3} in comparison with 2.13×10^{-2} and 2.17×10^{-2} at 60 and 65 °C, respectively. The half-life ($t_{1/2}$) of the enzyme extract at 55 °C was 315 min in comparison with 32.5 and 31.9 min for 60 and 65 °C, respectively (Table 2). Since the enzymatic activity of the crude GA extract at 55 °C was close to the activity at 60 and 65 °C, but more stable, it was adopted as the optimal digestion temperature for the subsequent food hydrolysis experiment.

$$\ln(E_t/E_0) = -k_d t \qquad (1)$$

$$t_{1/2} = \ln(2)/k_d \qquad (2)$$

TABLE 2: Deactivation constant (k_d) and half-lives ($t_{1/2}$) of the crude GA extract at 55, 60 and 65°C at pH 5.5.

Temperature (°C)	kd (minutes^{-1})	$t_{1/2}$ (minutes)
55	2.20×10^{-3}	315.0
60	2.13×10^{-2}	32.5
65	2.17×10^{-2}	31.9

4.2.4 APPLICATION OF CRUDE GLUCOAMYLASE EXTRACT ON MIXED FOOD WASTE HYDROLYSIS FOR GLUCOSE PRODUCTION

In the reality, mixed food waste is rich in salt [21] that may inhibit the enzymatic hydrolysis. To verify if the crude GA extract produced was applicable to food waste digestion for glucose production, it was used to hydrolyse the food waste which was collected from a local restaurant, under its optimal digestion conditions (at pH 5.5 and 55 °C). Increasing concentration of enzyme was added to the food waste and the time required for hydrolysis to produce glucose was determined (Figure 4). Significant difference in glucose concentration was only observed for the first hour but not after. In all cases, the food waste hydrolysis by the enzyme extracts was completed in 1 h. At the end of the hydrolysis, approximately 12 g/L glucose was produced and that was corresponding to approximately 53 g glucose produced from 100 g of mixed food waste (d.b.), while no production of glucose occurred in a control without crude GA extract. The amount of glucose produced from 100 g of mixed food waste hydrolysis was consistent with our previous study using a different hydrolysis approach [5].

The two commonly used approaches for cereal-based waste and food waste hydrolysis to produce glucose rich solution, include simultaneous fungus culturing and hydrolysis with the enzymes actively secreted [5,8,16,22,23] or direct addition of enzyme solution to digest the substrate [24]. In the first case, food waste hydrolysis is usually completed after 24 h [5,8,16,22,23]. In this study, the latter approach was adopted. Food waste hydrolysis by the crude GA extract was completed in 1 h under optimal conditions found. Similar experiment has been reported by Yan et al. using commercial GA for food waste hydrolysis. In their studies, food waste hydrolysis was completed in 2.5 h when GA to substrate ratio reached 80–140 U/g food waste [24]. When the GA to substrate ratio in the solution is reduced to 7.4 U/g substrate, 24 h was needed for complete substrate hydrolysis [16]. The higher efficiency for food waste hydrolysis (in 1 h) in this study was likely due to the higher initial GA to substrate ratio.

4.2.5 MATERIAL BALANCE FOR GLUCOSE PRODUCTION FROM 1 KG MIXED FOOD WASTE WITH CRUDE GLUCOAMYLASE EXTRACT

Scheme I shows the material balance of the studied process for crude GA production from pastry waste and glucose recovery from 1 kg mixed food waste (d.b.). All the calculations are provided in the supplementary information. According to our study, hydrolysis of 1 kg mixed food waste (d.b.) could lead to 0.53 kg glucose production. Approximately, 1.4 kg pastry waste (d.b.) is required to produce sufficient amount of crude GA extract for 1 kg mixed food waste hydrolysis. Furthermore, the material balance only presents the theoretical values of the up-scaled process based on our laboratory-scale experimental data. However, up-scale study is needed in order to demonstrate the process can be applied at industrial scale.

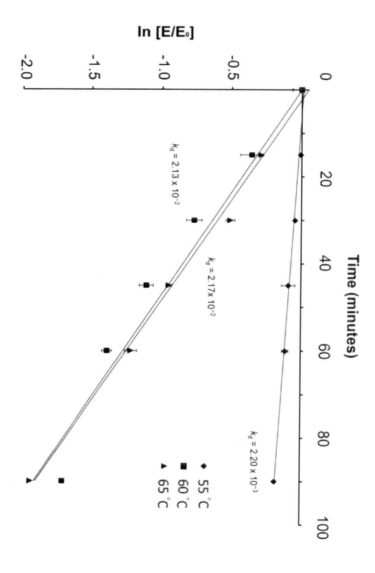

FIGURE 3: Thermal deactivation of crude GA extract at 55°C, (■) 60°C and (▲) 65°C over 90 min at pH 5.5. Experiments were duplicated. The mean values are plotted and the standard errors were reported.

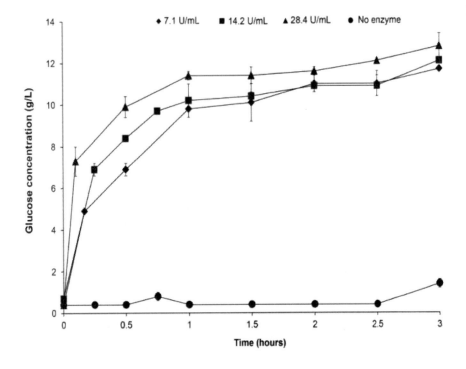

FIGURE 4: Hydrolysis of mixed food waste for glucose production in the presence of crude GA extract with (●) no enzyme, 7.1 U/mL, (■) 14.2 U/mL and (▲) 28.4 U/mL at pH 5.5 and 55°C for 3 h. Experiments were duplicated. The mean values are plotted and the standard errors are reported.

Production of Fungal Glucoamylase for Glucose Production

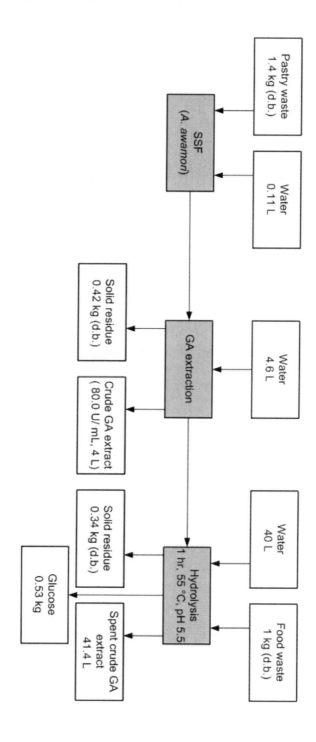

SCHEME 1: Material balance of the process in this study for processing 1 kg mixed food waste to produce glucose using crude GA extract produced from pastry waste based on the laboratory-scale experimental data.

4.3 EXPERIMENTAL SECTION

4.3.1 MICROORGANISM

Frozen spores of the fungus *Aspergillus awamori* (ATCC 14331), were used for SSF. Spores were suspended in demineralized water and loaded into conical flasks containing cornmeal agar and incubated at 30 °C for 7 days. Harvesting of fresh spores was carried out using 10% glycerol and the number of spores was counted using a haemocytometer. Fresh spore suspension of 1 mL was diluted to the required concentration with sterile demineralized water and used immediately for SSF.

4.3.2 FOOD WASTES PREPARATION

Pastry waste and mixed food waste were obtained from a local Starbucks store and canteen for SSFs and enzymatic hydrolysis experiments, respectively. Once the pastry waste was collected, it was homogenized with a kitchen blender and stored at −20 °C. Pastry waste was autoclaved before subjected to SSF and mixed food waste was lyophilized before enzymatic hydrolysis.

4.3.3 SOLID STATE FERMENTATION FOR GLUCOAMYLASE PRODUCTION

Sterilized pastry waste of 20 g was placed in a petri dish and inoculated with 1 mL of fresh spore suspension of *Aspergillus awamori* loaded onto the surface of the substrate. For each g of substrate, 5×10^5 spores were used. The plates were incubated under static condition at 30 °C for 11 days. Whole plate content was withdrawn regularly and analyzed for GA production.

4.3.4 GLUCOAMYLASE EXTRACTION

Whole content of the fermented solid in the petri dish was mixed thoroughly in a kitchen blender containing 20 mL demineralized water and the mixture was transferred into a 500 mL Duran bottle. The kitchen blender was rinsed with another portion of 20 mL demineralized water and pooled with the previously obtained mixture. The suspension was then mixed for 30 min with a magnetic stirrer at 30 °C, followed by centrifugation at 22,000 × g for 10 min. The supernatant was collected and filtered with Whitman no. 1 filter paper. The solution obtained refers to crude GA extract.

4.3.5 FOOD WASTE HYDROLYSIS FOR GLUCOSE PRODUCTION

Food waste hydrolysis was conducted by adding increasing amount of crude GA extract to test tubes containing 50 mg dried food waste in 2.2 mL of 0.2 M sodium acetate (pH 5.5). The reaction mixture was incubated at 55 °C for 3 h and mixed by pipetting for every 30 min. Aliquots of samples were withdrawn regularly from the reaction mixture and mixed with 10% (w/v) trichloroacetic acid prior to glucose determination.

4.3.6 GLUCOAMYLASE ACTIVITY

Activity of the crude GA extract was determined using the method described by Melikoglu et al. [17]. Wheat flour solution of 6% (w/v) was used as substrate and it was prepared in sodium acetate (pH 5.5) and gelatinized at 80 °C for 15 min before usage. The assay was conducted by mixing 0.25 mL of 5 times diluted crude GA extract and 0.5 mL of gelatinized wheat flour solution, and incubated at 55 °C. The reaction was terminated after 10 min by adding 0.25 mL 10% (w/v) trichloroacetic acid solution to the reaction mixture. The reaction mixture was centrifuged and the glucose concentration in the supernatant was analyzed using the Analox GL6 glucose analyzer. One unit (U) of GA activity is defined as the amount of enzyme that releases 1 µmol of glucose per minute under assay conditions and is expressed as U/g of dry substrate as described [12].

For determination of optimal reaction pH, the crude GA extract and the gelatinized wheat flour solutions were prepared in sodium acetate buffer from pH 3.5 to 7.5. Optimal reaction temperature determination of the crude GA extract was carried out at temperature range of 40–75 °C with 5 °C increment at optimal pH. Thermo-stability of the crude GA extract was investigated at different temperatures (55–65 °C) at optimal pH. Experiments were performed in duplicate and the standard errors were reported.

4.4 CONCLUSIONS

In this work, we have demonstrated the feasibility of pastry waste as feedstock for GA production. High GA yield (253.7 ± 20.4 U/g) of the crude enzyme extract was obtained without addition of nitrogen in comparison to other reported waste substrates, highlighting its potential as a feedstock for GA production in industrial scale. Under the optimal digestion conditions (pH 5.5 and 55 °C), the crude GA extract could hydrolyze mixed food waste in 1 h and generate around 53 g glucose from 100 g of mixed food waste. The work is of great significance as it shows sustainable GA production from food waste for potential municipal food waste treatment and sustainable chemicals production.

REFERENCES

1. Hong Kong SAR Environmental Protection Department. Monitoring of solid waste in Hong Kong – waste statistics for 2011, in Hong Kong 2011. Available online: https://www.wastereduction.gov.hk/en/materials/info/msw2011.pdf (accessed on 21 June 2013).
2. Sauer, J.; Sigurskjold, B.W.; Christensen, U.; Frandsen, T.P.; Mirgorodskaya, E.; Harrison, M.; Roepstorff, P.; Svensson, B. Glucoamylase: Structure/Function relationships, and protein engineering. Biochim. Biophys. Acta 2000, 1543, 275–293.
3. Norouzian, D.; Akbarzadeh, A.; Scharer, J.M.; Young, M.M. Fungal glucoamylases. Biotechnol. Adv. 2006, 24, 80–85.
4. Wang, X.Q.; Wang, Q.H.; Liu, Y.Y.; Ma, H.Z. On-Site production of crude glucoamylase for kitchen waste hydrolysis. Waste Manag. Res. 2010, 28, 539–544.
5. Pleissner, D.; Lam, W.C.; Sun, Z.; Lin, C.S.K. Food waste as nutrient source in heterotrophic microalgae cultivation. Bioresour. Technol. 2013, 137, 139–146.
6. Sayeki, M.; Kitagawa, T.; Matsumoto, M.; Nishiyama, A.; Miyoshi, K.; Mochizuki, M.; Takasu, A.; Abe, A. Chemical composition and energy value of dried meal from food waste as feedstuff in swine and cattle. Anim. Sci. J. 2001, 72, 34–40.
7. Koutinas, A.A.; Wang, R.; Webb, C. Estimation of fungal growth in complex, heterogeneous culture. Biochem. Eng. J. 2003, 14, 93–100.

8. Zhang, A.Y.-Z.; Sun, Z.; Leung, C.C.J.; Han, W.; Lau, K.Y.; Li, M.; Lin, C.S.K. Valorisation of bakery waste for succinic acid production. Green Chem. 2013, 15, 690–695.
9. Zambare, V. Solid state fermentation of Aspergillus. oryzae for glucoamylase production on agro residues. Int. J. Life Sci. 2010, 4, 16–25.
10. Ganzlin, M.; Rinas, U. In-Depth analysis of the Aspergillus. niger glucoamylase (glaA) promoter performance using high-throughput screening and controlled bioreactor cultivation techniques. J. Biotechnol. 2008, 135, 266–271.
11. Ventura, L.; González-Candelas, L.; Pérez-Gonzáez, J.A.; Ramón, D. Molecular cloning and transcriptional analysis of the Aspergillus. terreus gla1 gene encoding a glucoamylase. Appl. Environ. Microbiol. 1995, 61, 399–402.
12. Anto, H.; Trivedi, U.B.; Patel, K.C. Glucoamylase production by solid-state fermentation using rice flake manufacturing waste products as substrate. Bioresour. Technol. 2006, 97, 1161–1166.
13. Wang, Q.; Wang, X.; Wang, X.; Ma, H. Glucoamylase production from food waste by Aspergillus. niger under submerged fermentation. Process. Biochem. 2008, 43, 280–286.
14. Kareem, S.O.; Akpan, I.; Oduntan, S.B. Cowpea waste: A novel substrate for solid state production of amylase by Aspergillus. oryzae. Afr. J. Microbiol. Res. 2009, 3, 974–977.
15. Du, C.; Lin, S.K.C.; Koutinas, A.; Wang, R.; Dorado, P.; Webb, C. A wheat biorefining strategy based on solid-state fermentation for fermentative production of succinic acid. Bioresour. Technol. 2008, 99, 8310–8315.
16. Wang, R.; Godoy, L.C.; Shaarani, S.M.; Melikoglu, M.; Koutinas, A.; Webb, C. Improving wheat flour hydrolysis by an enzyme mixture from solid state fungal fermentation. Enzyme Microb. Technol. 2009, 44, 223–228.
17. Melikoglu, M.; Lin, C.S.K.; Webb, C. Stepwise optimisation of enzyme production in solid state fermentation of waste bread pieces. Food Bioprod. Process. 2013. in press.
18. Lawton, J.M.; Doonan, S. Thermal inactivation and chaperonin-mediated renaturation of mitochondrial aspartate aminotransferase. Biochem. J. 1998, 334, 219–224.
19. Johannes, T.W.; Woodyer, R.D.; Zhao, H. Directed evolution of a thermostable phosphite dehydrogenase for NAD(P)H regeneration. Appl. Environ. Microbiol. 2005, 71, 5728–5734.
20. Allen, M.J.; Coutinho, P.M.; Ford, C.F. Stabilization of Aspergillus. awamori glucoamylase by proline substitution and combining stabilizing mutations. Protein Eng. 1998, 11, 783–788.
21. Myer, R.O.; Brendemuhl, J.H.; Johnson, D.D. Evaluation of dehydrated restaurant food waste products as feedstuffs for finishing pigs. J. Anim. Sci. 1999, 77, 685–692.
22. Leung, C.C.J.; Cheung, A.S.Y.; Zhang, A.Y.-Z.; Lam, K.F.; Lin, C.S.K. Utilisation of waste bread for fermentative succinic acid production. Biochem. Eng. J. 2012, 65, 10–15.
23. Dorado, M.P.; Lin, S.K.C.; Koutinas, A.; Du, C.; Wang, R.; Webb, C. Cereal-Based biorefinery development: Utilisation of wheat milling by-products for the production of succinic acid. J. Biotechnol. 2009, 143, 51–59.

24. Yan, S.; Yao, J.; Yao, L.; Zhi, Z.; Chen, X.; Wu, J. Fed batch enzymatic saccharification of food waste improves the sugar concentration in the hydrolysates and eventually the ethanol fermentation by Saccharomyces. Cerevisiae H058. Braz. Arch. Biol. Technol. 2012, 55, 183–192.

PART III

COMPOSTING

CHAPTER 5

Changes in Selected Hydrophobic Components During Composting of Municipal Solid Wastes

JAKUB BEKIER, JERZY DROZD, ELZBIETA JAMROZ,
BOGDAN JAROSZ, ANDRZEJ KOCOWICZ,
KAROLINA WALENCZAK, AND JERZY WEBER

5.1 INTRODUCTION

One of the most practical ways to utilise municipal solid waste is composting, thereby producing materials that may be productively used to improve soil properties (Weber et al. 2007). Municipal solid waste (MSW) have variable composition, with the main ingredients being: comestible waste products (40–60 %); paper and cardboard (20 %); and glass, plastic and metal up to 40 % (Costa et al. 1991). During composting, some of these materials are biotransformed into more stable products that are rich in humic substances. According to numerous studies, the quality and quantity of these humic materials influences the stability and maturity of the final product (Garcia et al. 1992; Spaccini and Piccolo 2009). However, the

---

Changes in Selected Hydrophobic Components During Composting of Municipal Solid Wastes. © Bekier J, Drozd J, Jamroz E, Jarosz B, Kocowicz A, Walenczak K, and Weber J. Journal of Soils and Sediments *14*,2 (2014). DOI: 10.1007/s11368-013-0696-0. *This article is licensed under a Creative Commons License, http://creativecommons.org/licenses/by/3.0/.*

actual structure of the humic substances remains controversial (de Leeuw and Largeau 1993). Most studies suggest a predominance of aromatic units in matured humic substances, while other results indicate largely aliphatic structures in humic extracts. These dissimilarities may result from differing microbial activities, especially in resynthesis processes occurring during composting (Gea et al. 2007).

Composted MSW may contain variable amounts of hydrophobic constituents, such as lipids, fats, waxes, resins, which in soil science are determined as bitumen fraction of humin. Many authors indicate these substances as an integral part of the organic matter present in municipal wastes (Amir et al. 2006; Gea et al. 2007; Réveillé et al. 2003; Veeken et al. 2000). Fats and oils originating from common households consist of straight-chain fatty acids in the form of glycerol esters (Ruggieri et al. 2008; Steger et al. 2003; Wakelin and Forster 1997). Fatty acids and lipids are integral constituents of cellular membranes, leftovers and cellular secretions, and participate in many biological processes (Klamer and Bååth 2004; Ryckeboer et al. 2003). The main sources of these substances in municipal wastes are food refuse and kitchen waste, which contain varying amounts of animal and plant fats (Komilis and Ham 2003; Maliki and Lai 2011). Due to difficulties in monitoring the transformation of lipids and fatty acids into humic substances, some authors consider the role of these substances in humification processes as mainly that of energy donors for microorganisms (Chen et al. 1997; Komilis and Ham 2003; Ryckeboer et al. 2003). Conversely, the results of other studies indicate that fatty acids play an integral part in the formation of humic substances and are important during the creation of humic acid aliphatic structures (Amir et al. 2006; Hachicha et al. 2009; Lguirati et al. 2005; Spaccini and Piccolo 2009). From this point of view, qualitative and quantitative changes in fatty acids during composting may be considered as chemical indices of compost maturity (Barje et al. 2008; Gea et al. 2007; Hachicha et al. 2009; Ruggieri et al. 2008).

As a result of its chemical properties, complicated structure and hydrophobic character, the biotransformation and decomposition of fats are dependent on several factors. The most important of these are microbiological activity, temperature, pH, humidity and the nature of the macro- and microelements (Amir et al. 2008; Barje et al. 2008; Gea et al. 2007). Many studies have reported the biotransformation of fats, although the intensity of their biodegradation is generally limited by unfavourable physiochemical properties, such as insolubility in water (Becker et al. 1999; Lefebvre et al. 1998). Routine composting technologies enable the biotransformation of wastes enriched with fats, providing these fats do not exceed 15 % of the dry mass (Filippi et al. 2002; Garcia-Gomez et al. 2003). However, this may cause an extended thermophilic phase, as a result of the high chemical energy of the fatty acids (Nakano and Matsumura 2001). Lipid biotransformation and degradation processes are usually most intensive in the thermophilic phase, during which time decreases of 80–90 % from initial values have been observed, with reductions of up to 97 % occurring during long composting periods (Baddi et al. 2004; Ruggieri et al. 2008). Biological treatment of fats under thermophilic conditions is considered to be the appropriate method for producing the required changes in the physical and chemical properties of these hydrophobic compounds (Hachicha et al. 2009; Gea et al. 2007). From this perspective, composting can be considered as an alternative method for the effective treatment of fats and oils.

Wastes, as well as mature compost, contain hydrophobic substances, including fats, which biodegradability is considered as largely limited due to unfavourable properties, such as insolubility in water (Gea et al. 2007; Lefebvre et al. 1998; Ruggieri et al. 2008). However, their biotransformation during composting processes may be influenced by other factors, mainly temperature, oxygenation and mineral composition (Barje et al. 2008). The aim of the present study was to determine qualitative and quantitative changes of hydrophobic substances, especially fatty acids, during the course of MSW composting. This provides new information on intensity of hydrophobic versus other substances decomposition undergoing during these processes.

5.2 MATERIALS AND METHODS

5.2.1 COMPOST SAMPLES

The presented research concerns material processed at the Katowice commercial composting plant (Upper Silesia, Poland), where randomly collected raw material was composted for 180 days in a pile, after initial biostabilisation according to MUT-DANO technology. Changes in temperature were measured each day throughout the composting process. Samples for analysis were taken after days 1, 14, 28, 42, 56, 90 and 180 of the composting from three different places of the pile and then averaged. The weight of each averaged samples was approximately 2.5–3.0 kg of fresh mass. All anthropogenic contaminations (glass, plastic, metal, etc.) were removed manually. The moisture was determined in collected material while for further determinations, samples were dried, grounded and sieved through a sieve with diameter 2.0 mm.

5.2.2 TOTAL ORGANIC CARBON, HYDROPHOBIC SUBSTANCES CARBON AND FATTY ACID CARBON

Total organic carbon (TOC) content was determined with automatic analyzer (Ströhlein CS-MAT 5500, Germany). Hydrophobic substances were extracted by means of an ethanol and benzene mixture (1:2 v/v) and fats were extracted with petroleum ether, using the Soxhlet extractor. After extraction, samples were dried in controlled condition at temperature 40 °C for 24 h to evaporate extractant. Hydrophobic substances carbon (HSC) was calculated as the difference in organic carbon before and after extraction and fatty acid carbon (FAC), determined with automatic analyzer (Ströhlein CS-MAT 5500, Germany).

5.2.3 GC ANALYSIS

To analyse qualitative and quantitative changes of fatty acids during compost maturation, catalysed transesterification with BF_3 in methanol method was used (Metcalfe and Schmitz 1961). In short, fats from petroleum ether extract were hydrolysed with 1 M NaOH in methanol and then esterified after addition of BF3 14 % solution in methanol. Obtained fatty acids methyl esters (FAMEs) were isolated with hexane and identified using an gas chromatograph (GC)/mass spectrometer (MS) apparatus (Agilent 6890N gas chromatograph coupled with 5973 MS detector), according to their mass spectra. After qualitative identification, the relative FAMEs content in each sample was assessed in GC using an Agilent 6890N gas chromatograph equipped with the flame ionization detector (FID). In the FID mode, the presence of the main FAMEs isomers was confirmed with the authentic samples of hexadecanoic (i.e. palmitic), octadecanoic (i.e. stearic), oleic, linoleic and linolenic methyl esters. Thus, the remaining minor FAMEs peaks with identical molecular ions were ascribed to the isomeric methyl esters, most likely belonging to iso and anteiso series typical for bacteria.

5.2.4 STATISTICAL ANALYSIS

TOC, HSC and FAC were determined by three repetitions for each sample. Mean values, presented in tables or on figures, were used to characterise the transformation processes. Statistical analyses presented by correlation coefficients between TOC, HSC, FAC and composting parameters were calculated on a basis of all obtained results using the 'Statistica 7' package at the $p < 0.05$ level (Table 1).

TABLE 1: Correlation coefficients between TOC, HSC, FAC and composting parameters

Parameter	TOC	HSC	FAC
Composting time	−0.97*	−0.84*	−0.70*
Temperature	−0.87*	−0.94*	−0.83*

*$p < 0.05$, correlation significance

5.3 RESULTS AND DISCUSSION

5.3.1 TOC, HSC, AND FAC TRANSFORMATION

The intensity of organic matter biotransformation and decomposition processes during composting is controlled by temperature, humidity, pH and the properties of the organic and mineral substances (Licznar et al. 2010), which affect microbiological activity in the composted material. Over the 180 days of the experiment, the amount of TOC decreased from 201.7 to 100.1 g kg^{-1} (Table 2), with the greatest intensity of this process being observed during the thermophilic composting phase (Fig. 1). This phase with average temperature >55 °C started on 20th day, the highest temperature (64 °C) was observed on 40th day of composting (see Fig. 1) and was prolonged until about the 60th day of the composting. The reason of a such long thermophilic phase was not sufficiently regular conversion of the material in the pile.

TABLE 2: Contents of total organic carbon (TOC), hydrophobic substances carbon (HSC) and fatty acids carbon (FAC) during the experiment

Composting time	TOC	HSC	FAC	Unsaturated	Saturated
Days		g kg^{-1}		% FAC	
1	201.7	27.8	1.98	61.9	31.5
14	192.0	21.2	1.32	54.9	37.6
28	184.9	17.7	0.76	55.0	39.0
42	165.0	13.5	0.41	46.0	46.3
56	157.4	9.3	0.34	45.7	53.8
90	132.8	5.8	0.13	39.2	56.8
180	100.1	4.5	0.17	28.7	64.8

The HSC content decreased from 27.8 to 5.8 g kg^{-1} during the first 90 days of composting (see Table 2). After that time, it remained at the same level, due to the cooling and stabilisation of the composting material in the mesophilic phase (Chefetz et al. 1996).

More intensive changes were observed in FAC, which decreased from 1.98 to 0.13 g kg^{-1} during the first 90 days (see Table 2). After that period, the fatty acid content increased slightly, presumably as a result of stabilisation and changes in microbial composition, as suggested by Amir et al. (2008) and Ryckeboer et al. (2003). Obtained results indicate that hydrophobic substances, especially fatty acids, are transformed more intensively than other compounds of composted MSW material.

Changes in Selected Hydrophobic Components

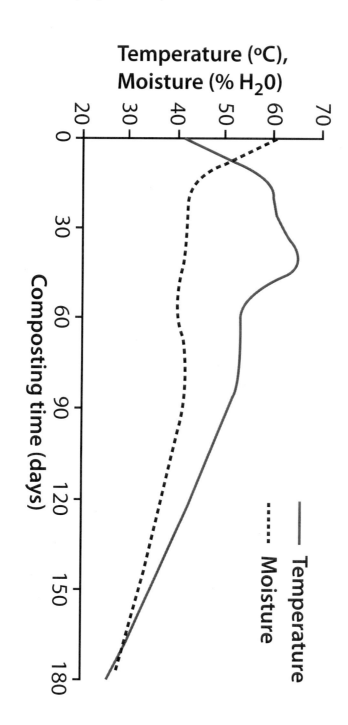

FIGURE 1: Changes in temperature and humidity during composting of municipal solids wastes

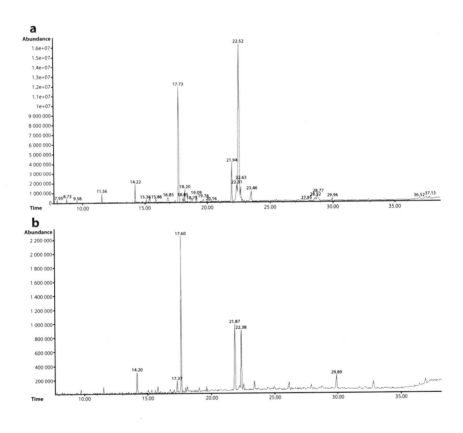

FIGURE 2: Fatty acids in the capillary GC of extracted lipids at a the beginning of the experiment and b after 180 days of composting

5.3.2 FATTY ACID TRANSFORMATION

During the course of the whole experiment, 24 fatty acids were identified, of which 14 were present only in trace amounts (Fig. 2). Consequently, only 10 fatty acids (six saturated and four unsaturated) were selected for detailed investigation (Tables 3 and 4).

TABLE 3: Changes in the relative amounts of unsaturated fatty acids (percent of total fatty acids) during composting

Composting time	Number of carbon atoms/number of C=C bonds			
Days	16:1	18:1	18:2	20:1
1	1.0	56.1	2.3	2.5
14	1.1	49.8	2.6	1.4
28	1.1	49.9	2.8	1.2
42	1.2	40.8	2.5	1.5
56	1.4	40.8	2.1	1.4
90	1.7	32.5	3.6	1.4
180	2.1	23.0	2.7	0.9

TABLE 4: Changes in the relative amounts of saturated fatty acids (percent of total fatty acids) during composting

Composting time	Number of carbon atoms/number of C=C bonds					
Days	15:0	16:0	17:0	18:0	20:0	22:0
1	1.2	18.8	1.9	8.3	0.8	0.5
14	1.5	22.3	2.1	10.3	0.9	0.8
28	1.2	24.0	1.7	10.5	0.8	0.8
42	1.5	26.5	2.2	14.0	1.1	1.0
56	2.4	30.2	2.2	15.7	1.5	1.8
90	2.3	30.2	2.3	18.9	1.3	1.8
180	2.7	36.7	2.3	19.4	1.5	2.2

Unsaturated fatty acids dominated in the raw material (see Table 2), which is typical for households. Octadecenoic acid (18:1) is the main lipidic component of almost all cellular membranes occurring in living organisms, and it was the main unsaturated fatty acid. Its share reached 56 % of the FAC in the raw material, and decreased to 23 % after 180 days of composting processes (see Table 3). The biggest changes were observed during the first 40 days, which correspond to the declining number of microorganisms during compost maturation (Hachicha et al. 2009; Klamer and Bååth 2004).

Despite the fact that composting processes generally lead to a decrease in FAC (see Table 2), individual fatty acids indicated varying transformation intensities. Especially during the first few weeks of composting, hexadecanoic (16:0) and octadecanoic (18:0) acids increased from 18.8 to 36.7 % and from 8.3 to 19.4 % of the total FAC, respectively (see Table 4). These acids, known as long-chain fatty acids (LCFAs), are integral components of cell membranes and usually occur in large amounts in living organisms and composted materials (Gea et al. 2007; Klamer and Bååth 2004; Ruggieri et al. 2008). The obtained results need further investigation in order to understand the release mechanisms of the various LCFAs from their parent glycerolipids, and their transformations during composting.

The role of abiotic processes in the transformation of lipid substances present in the raw material also merits further enquiry. In the presence of oxygen, unsaturated fatty acids in composted material undergo autoxidation, producing a series of hydroperoxides. These subsequently stimulate decomposition, thus contributing to the release of volatile odorous compounds, including hydrocarbons, saturated and unsaturated aldehydes, alcohols and some other oxo compounds (Frankel 1980). Similar volatile patterns were observed in the microbial decomposition of fatty foods (Ercolini et al. 2009). It is assumed that at elevated temperatures, and especially in the presence of metal ions (which act as catalysts), the autoxidation reactions must have been more intense, although the occurrence of these odorous substances was not the subject of this study. The presence of lignin-modifying basidiomycetous fungi may be considered as an additional factor influencing the formation of peroxides (Kapich et al. 2011).

In any case, both biotic and abiotic processes justify the decrease in fatty acids during the composting process. It is worth noting that, according to the results of other researchers, fungi isolated from compost do not produce these volatiles in the absence of lipid substances (Fischer et al. 1999). The observed changes and tendencies are consistent with the results of Amir et al. (2008), Gea et al. (2007) and Ruggieri et al. (2008). In the final stage of the composting processes, the fatty acids were dominated by saturated forms, primarily hexadecanoic (16:0) acid. The share of this component increased from 18.8 % at the beginning of the experiment to 36.7 % of FAC after 180 days (see Table 4).

Within the composted material some particular fatty acids with odd numbers of carbon atoms, specifically pentadecanoic (15:0) and heptadecanoic (17:0), were present, which is a characteristic of many bacteria (see Table 4). These compounds can be synthesised by bacteria de novo, or may be products of lipid matter transformation, mainly during the thermophilic phase (Řezanka and Sigler 2009). This confirms that high microbiological activity is caused by thermo-basophilic bacteria (Amir et al. 2008; Ryckeboer et al. 2003), which are therefore responsible for most processes associated with the biotransformation of organic matter, including lipid substances, and especially fatty acids.

Figure 2 presents capillary gas chromatograms of FAMEs obtained from lipids extracted during various phases of the composting process. Those peaks with retention times equal to approximately 17 min correspond to saturated acids, while those of 21–22 min correspond to unsaturated fatty acids. This confirms that during the composting and maturation processes, the ratio of saturated vs. unsaturated fatty acids increases in comparison with the fresh compost.

Since the total amount of fatty acids clearly decreased during composting, the increased proportion of saturated fatty acids represents their greater resistance to decomposition than unsaturated forms. The amount of FAC slightly increased at the end of the composting, in comparison to that present after 90 days (see Table 2). Intensity of the transformation processes decrease during maturation stage, due to lowering of microbiological activity, what is typical for the stabilisation phase of compost production (Chefetz et al. 1996; Ryckeboer et al. 2003), indicating that fungi, actinomycetes, mesophilic bacteria and other microorganisms—that dominate the microbial community—are unable to transform the fatty acids.

5.4 CONCLUSIONS

The extent of decomposition of hydrophobic substances, especially fatty acids, is greater than other components of composted municipal solid waste, and intensity of the biotransformation is significantly correlated with composting parameters, mainly temperature and time. During the thermophilic phase of municipal solid waste composting, the decrease in total content of hydrophobic substances is approximately fivefold, while the reduction in fatty acids can be about tenfold. Unsaturated fatty acids are more intensively decomposed during the composting processes, while saturated fatty acids are more resistant. Moreover, transformation of fatty matter may result in the creation of specific isomers with odd numbers of carbon atoms.

REFERENCES

1. Amir S, Hafidi M, Lemee L, Merlina G, Guiresse M, Pinelli E, Revel J-C, Bailly J-R, Ambles A (2006) Structural characterization of humic acids, extracted from sewage sludge during composting, by thermochemolysis—gas chromatography—mass spectrometry. Process Biochem 41(2):410–422
2. Amir S, Merlina G, Pinelli E, Winterton P, Revel C, Hafidi M (2008) Microbial community dynamics during composting of sewage sludge and straw studied through phospholipid and neutral lipid analysis. J Hazard Mater 159(2–3):593–601
3. Baddi GA, Alburquerque JA, Gonzálvez J, Cegarra J, Hafidi M (2004) Chemical and spectroscopic analyses of organic matter transformations during composting of olive mill wastes. Int Biodeter Biodegr 54(1):39–44
4. Barje F, Amir S, Winterton P, Pinelli E, Merlina G, Cegarra J, Revel J-C, Hafidi M (2008) Phospholipid fatty acids analysis to monitor the co-composting process of olive oil mill wastes and organic household refuse. J Hazard Mater 154(1–3):682–987
5. Becker P, Koster D, Popov MN, Markossian S, Antranikian G, Markl H (1999) The biodegradation of olive oil and the treatment of lipid-rich wool scouring wastewater under aerobic thermophilic conditions. Water Res 33(3):653–660
6. Chefetz B, Thatcher PG, Hadar Y, Chen Y (1996) Chemical and biological characterization of organic matter during composting of municipal solid waste. J Environ Qual 25(4):776–785

7. Chen Y, Chefetz B, Adani F, Genevini P, Hadar Y (1997) Organic matter transformation during composting of municipal solid waste. In: Drozd J, Gonet SS, Senesi N, Weber J (eds) The role of humic substances in ecosystems and in environmental protection. PTSH, Wroclaw, pp 795–804
8. Costa F, Garcia C, Hernandez T, Polo A (1991) Residous organicos urbanos. Manejo y utilization. CSIC, Murcia
9. de Leeuw JW, Largeau C (1993) A review of macromolecular organic compounds that comprise living organisms and their role in kerogen, coal and petroleum formation. In: Engel MH, Macko SA (eds) Organic geochemistry, principles and applications. Plenum, New York, pp 23–72
10. Ercolini D, Russo F, Nasi A, Ferranti P, Villani F (2009) Mesophilic and psychrotrophic bacteria from meat and their spoilage potential in vitro and in beef. Appl Environ Microb 75(7):1990–2001
11. Filippi C, Benidi S, Levi-Minzi R, Cardelli R, Saviozzi A (2002) Co-composting of olive oil mill by-products: chemical and microbiological evaluations. Compost Sci Util 10(1):63–71
12. Fischer G, Schwalbe R, Möller M, Ostrowski R, Dott W (1999) Species-specific production of microbial volatile organic compounds (MVOC) by airborne fungi from a compost facility. Chemosphere 39(5):795–810
13. Frankel EN (1980) Lipid oxidation. Prog Lipid Res 19:1–22
14. Garcia C, Hernandez T, Costa F (1992) Characterization of humic acids from uncomposted and composted sewage sludge by degradative and nondegradative techniques. Bioresour Technol 41(1):53–57
15. Garcia-Gomez A, Roig A, Bernal MP (2003) Composting of the solid fraction of olive mill wastewater with olive leaves: organic matter degradation and biological activity. Bioresour Technol 86(1):59–64
16. Gea T, Ferrer P, Alvaro G, Valero F, Artola A, Sanchez A (2007) Co-composting of sewage sludge: fats mixtures and characteristics of the lipases involved. Biochem Eng J 33(3):275–283
17. Hachicha R, Hachicha S, Trabelsi I, Woodward S, Mechichi T (2009) Evolution of the fatty fraction during co-composting of olive oil industry wastes with animal manure: maturity assessment of the end product. Chemosphere 75(10):1382–1386
18. Kapich AN, Korneichik TV, Hammel KE, Hatakka A (2011) Comparative evaluation of manganese peroxidase- and Mn(III)-initiated peroxidation of C18 unsaturated fatty acids by different methods. Enzyme Microb Tech 49(1):25–29
19. Klamer M, Bååth E (2004) Estimation of conversion factors for fungal biomass determination in compost using ergosterol and PLFA 18:2ω6,9. Soil Biol Biochem 36:57–65
20. Komilis D, Ham RK (2003) The effect of lignin and sugars to the aerobic decomposition of solid waste. Waste Manage 23(5):419–423
21. Lefebvre X, Paul E, Mauret M, Baptiste P, Capdeville B (1998) Kinetic characterization of saponified domestic lipid residues aerobic biodegradation. Water Res 32(10):3031–3038
22. Lguirati A, Baddi G, El Mousadik A, Gilard V, Revel JC, Hafidi M (2005) Analysis of humic acids from aerated and non-aerated urban landfill compost. Int Biodeter Biodegr 56(1):8–16

23. Licznar M, Drozd J, Licznar SE, Weber J, Bekier J, Walenczak K (2010) Effect of municipal waste moisture level on transformations of nitrogen forms in the course of composting. Ecol Chem Eng A 17(7):787–797
24. Maliki AD, Lai K-M (2011) Design and application of a pre-composting test step to determine the effect of high fat food wastes on an industrial scale in-vessel composting system. Int Biodeter Biodegr 65(6):906–911
25. Metcalfe LD, Schmitz AA (1961) The rapid preparation of fatty acid esters for gas chromatographic analysis. Anal Chem 33(3):363–364
26. Nakano K, Matsumura M (2001) Improvement of treatment efficiency of thermophilic oxic process for highly concentrated lipid wastes by nutrient supplementation. J Biosci Bioeng 92(6):532–538
27. Réveillé V, Mancuy L, Jardé E, Garnier-Sillan E (2003) Characterization of sewage sludge derived organic matter: lipids and humic acids. Org Geochem 34(4):615–627
28. Řezanka T, Sigler K (2009) Odd-numbered very-long-chain fatty acids from the microbial, animal and plant kingdoms. Prog Lipid Res 48(3–4):206–238
29. Ruggieri L, Artola A, Gea T, Sanchez A (2008) Biodegradation of animal fats in a co-composting process with wastewater sludge. Int Biodeter Biodegr 62(3):297–303
30. Ryckeboer J, Mergaert J, Coosemans J, Deprins K, Swings J (2003) Microbiological aspects of biowaste during composting in a monitored compost bin. J Appl Microbiol 94(1):127–137
31. Spaccini R, Piccolo A (2009) Molecular characteristics of humic acids extracted from compost at increasing maturity stages. Soil Biol Biochem 41:1164–1172
32. Steger K, Asa J, Sven S, Ingvar S (2003) Comparison of signature lipid methods to determine microbial community structure in compost. J Microbiol Meth 55(2):371–382
33. Veeken A, Nierop K, de Wilde V, Hamelers B (2000) Characterization of NaOH-extracted humic acids during composting of a biowaste. Bioresour Technol 72(1):33–41
34. Wakelin NG, Forster CF (1997) An investigation into microbial removal of fats, oils and greases. Bioresour Technol 59(1):37–43
35. Weber J, Karczewska A, Drozd J, Licznar M, Licznar SE, Jamroz E, Kocowicz A (2007) Agricultural and ecological aspects of a sandy soil as affected by the application of municipal solid waste compost. Soil Biol Biochem 39:1294–1302

CHAPTER 6

Transforming Municipal Waste into a Valuable Soil Conditioner through Knowledge-Based Resource-Recovery Management

GOLABI MH, KIRK JOHNSON, TAKESHI FUJIWARA, AND ERI ITO

6.1 INTRODUCTION

Rapid increases in the volume and variety of solid and hazardous waste as a result of continuous economic growth, urbanization, and industrialization are a burgeoning problem for national and local governments, which must ensure effective and sustainable management of waste [1]. Between 2007 and 2011, global generation of municipal waste has been estimated to have risen by 37.3%, equivalent to roughly an 8% increase per year [1]. The EU has estimated that its 25 member states produce 700 million tons of agricultural waste annually [1].

As reported by the United Nations Environmental Program [1], developing countries face difficult challenges to proper management of their waste; most effort is devoted to reducing the final volume and to generating sufficient funds for waste management. If most of the waste could be diverted through material and resource recovery, then a substantial reduction in final volumes of waste could be achieved, and the recovered material and resources could be used to generate revenue to fund waste management. This scenario forms the premise for Integrated Solid Waste Management; a system based on the 3R (Reduce-Reuse-Recycle) principle [1].

Appropriate segregation and recycling systems have been shown to divert significant quantities of waste from landfills and to convert them into resources [2]. On Guam, over the past several decades, solid-waste generation and disposal have transitioned from a concern needing a remedy to a crisis of monumental proportions.

Although Guam is a small, isolated tropical island with a population of just over 160,000 people, the island generates more than 90,000 tons of waste material each year [3]. The need for a comprehensive solid-waste management and recycling plan is therefore urgent if Guam is to minimize cost and avoid the undesirable environmental effects of legal and illegal dump sites. A comprehensive waste management would also allow for the use of recyclable as well as green and other organic refuse that is currently discarded in landfills as sources of producing organic soil conditioner for a sustainable agricultural cropping system in Guam and the other island in the Micronesian region.

Waste reduction and recycling are fundamental to any future waste-management strategy on Guam and other islands of Micronesia. Accurate information on waste generation, especially waste characteristics, is also needed for study of the feasibility of such strategies on Guam and its neighbors. Unfortunately, presently available data on Guam [4,5] are not reliable enough for development of a comprehensive management and recycling strategies, and information is lacking on social behavior and life style that may strongly affect the character and production of waste.

Basic data from the unregulated local Ordot landfill over the past several years suggest that Guam residents produce on average more waste per capita than the rest of the United States [6]. Because any waste-management policy must find meaning and purpose within the framework of consumption patterns if it is to be effective at all. Residents must come to know and understand what, how much, and more importantly why they consume and hence must become aware of the impact that such consumption has on their island's environment and economy and on the social and cultural life of their community.

Residents' education about awareness of the types and amounts of waste generated and its handling are essential parts of any waste-management strategy that would be economically feasible, culturally acceptable, and environmentally sustainable while maintaining the integrity of the island's natural resources.

6.1.1 SURVEY QUESTIONNAIRE FOR EDUCATIONAL PURPOSES

To collect the necessary data, we have developed and used questionnaires as a surveying tool. The results are expected to help us understand the social behavior and the residents' life style as the first step toward the development of a sound and effective waste-management strategy for the island of Guam.

To determine not only the composition of waste by components but also citizens' consciousness and knowledge about waste reduction and recycling, we developed a citizen questionnaire designed for statistical analysis not only as a management tool but also as a way of educating the general public. The contents of the survey questionnaire included (1) educational background of head of the household, (2) public awareness of environmental problems associated with waste, (3) waste characteristics and willingness to segregate its components, and (4) participation in reduce-reuse-recycle activities.

By analyzing survey results, we hoped to develop a model of waste generation and citizens' consciousness of reduce-reuse-recycle principles for use in a comprehensive waste-management strategy. We expect that the survey results representing a true sample of the citizens of Guam will aid in the distribution and processing (collecting, compiling) of waste necessary to obtain high recovery ratios.

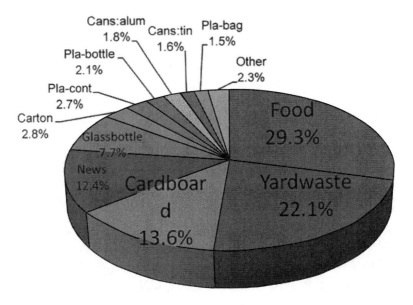

FIGURE 1: Recyclable waste produced by a Guam household. Placont, plastic containers; pla-bottle, plastic bottles; Pla-bag, plastic bags; News, Newspapers; alum, aluminum cans.

FIGURE 2: An Isfahan compost factory waste-gathering and transportation vehicle.

FIGURE 3: A large front-end loader pushes the garbage to the conveyors, immediately and cleaning the drop-off site.

FIGURE 4: A hammer mill equipped with double rotors and antiexplosives, which is used for breaking up large pieces.

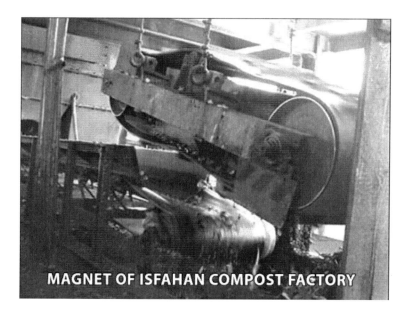

FIGURE 5: Magnets used to separate metals form the remainder of the waste stream.

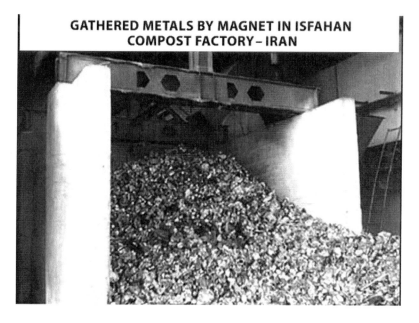

FIGURE 6: The metals trapped by the magnets fall into a separate compartment for transfer to other sections of the plant for further processing.

6.1.2 DETERMINATION OF THE WASTE GENERATION AND CHARACTERIZATION

Although reports by the European Commission [7] and studies by Martinho and Silveira [8] recommend sampling of waste containers placed in public areas (e.g., apartment complexes) as an ideal sampling technique, doing so would entail higher costs than did our survey.

Our approach not only obtained up-to-date data but also contributed to education of the public about waste management while promoting "zero waste." It also educated the public, private sector, as well as government agencies about composting and recycling of large-scale organic wastes. The survey approach has revealed that up to 77% of household waste on Guam is organic (food stuffs, yard wastes, newspapers, etc., Figure 1) in nature. This humongous amounts of wastes generated in a small island could easily be recycled through large-scale composting, as it is done at the Isfahan 'composting factory' described below (Figures 2–10). The remaining 23% of the nonorganic waste material like plastic bottles, cans, durable goods, etc., could also be recycled, leading to a "zero waste" management strategy that might require no land-filling cost for the community.

We therefore introduce here the idea of zero-waste management by presenting an example adopted in the city of Isfahan, Iran. We hope that presentation of such an example will lead local government leaders and the private sectors on Guam and the other islands of Micronesia, as well as major cities around the world, to consider adopting such a strategy.

6.2 ZERO WASTE MANAGEMENT STRATEGIES

Different zero-waste management strategies and techniques have been developed and adopted in different countries [9], but the strategy used in the city of Isfahan, Iran, which includes large-scale mechanical composting as a major component is of particular interest to Guam and the neighboring islands in the Micronesian region. The Isfahan composting operation is not only a recycling facility but also an organic fertilizer production plant that uses the organic waste generated by the city of Isfahan as a major source for its production lines.

FIGURE 7: Drum sieves, used to screen garbage through mesh into different size classes for further screening on vibrating conveyors.

FIGURE 8: Organic materials being conveyed to composting windrows for the fermentation process.

FIGURE 9: Composting windrow during the fermentation and turning process.

FIGURE 10: Mixing and turning of organic waste by a compost turner, which turns each windrow every four days.

Transforming Municipal Waste into a Valuable Soil Conditioner

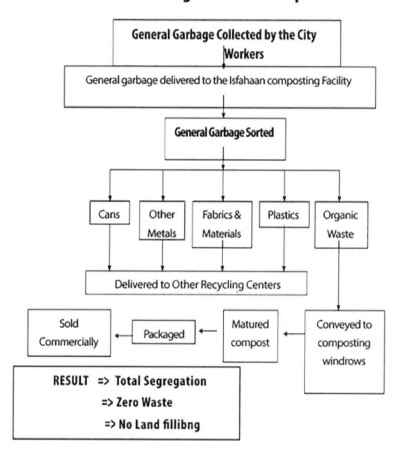

FIGURE 11: Chart showing the steps from waste collection to final compost production.

6.2.1 ISFAHAN COMPOST FACILITY AND RECYCLING TECHNIQUES

The Isfahan compost facility has two production lines, each able to handle 750 tons of garbage per day. Garbage is gathered at night at the transfer stations and sent to the recycling facility in both small and large hauling trucks (Figure 2). Upon delivery, the garbage is pushed to the conveyors by front-end loaders, and the receiving station is cleared as soon as trucks leave the site (Figure 3). Conveyors then carry the trash to vibrating screens that separate plastics, glass, cloth, etc. The trash is then carried on the conveyors to hammer mills, equipped with double rotors and anti-explosives, which break up the large pieces included in the bulk of trash (Figure 4).

A second set of vibrators is used to unpack organic materials and loosen everything else. Next, two magnets separate metals from the rest (Figure 5). The metals, aluminum cans, etc., all fall into a separate compartment (Figure 6) and are transferred to the other section of the facility for further processing and/or recycling/packaging. The remaining garbage is then carried by vibrating conveyors to drum sieves for screening through different mesh sizes and for further screening (Figure 7).

6.2.2 ADDITIONAL SORTING

Additional drums further screen plastics, glass bottles, wood, cloth, paper, etc. At this stage, items like fabric and cloth are separated, placed in a special compartment, and hauled away to be pressed into bales, which are sold for processing into pulp and other biodegradable material [9].

6.2.3 SEPARATION OF ORGANIC MATTER

Organic materials (technically waste containing carbon, including paper, plastics, wood, food wastes, and yard wastes; and other materials derived from plant or animal sources and decomposable by microorganisms [1] pass through drums of 50-mm mesh size and are then sent on separate

Transforming Municipal Waste into a Valuable Soil Conditioner 99

conveyors to fermentation sections (Figure 8), where it is placed in windrows and mixed for aeration every four days (Figures 9 and 10) until the material become a mature compost and ready for marketing after a comprehensive laboratory testing.

6.2.4 FACILITY PRODUCTION CAPACITY

Each year, from over 270,000 tons of input, the facility produces approximately 30,000 tons of fine mature compost, of which are sold to farmers, ranchers, horticulturists, and private gardeners as soil amendment/soil conditioner. Also, the facility produces about 12,000 tons of coarser compost as mulch, which is sold to municipalities for landscaping and for maintaining green spaces in many parks in Isfahan. The remaining biodegradable materials (fabrics, discarded cloth, etc.) are also sold to other cities within Iran as well as other neighboring countries for use in the pulp production and other similar manufacturing.

It is worth mentioning that all the products are tested at the factory's state-of-the-art laboratory for quality assurance before they are shipped to vendors and general customers and users of compost and mulch.

Figure 11 provides a flow-chart summary of the steps from waste collection to final compost production at the Isfahan composting factory.

6.2.5 RELEVANCE OF THE ISFAHAN COMPOSTING TECHNOLOGY FOR GUAM

The survey project described above served as a part of a comprehensive approach that includes increasing public awareness of comprehensive waste-management strategies for the island of Guam as well as the other islands in the western Pacific.

In addition to waste characterization by means of the survey questionnaires, presentation of the "Isfahan Waste Management System" also provides a knowledge-based foundation that we hope will lead to adaptation of the technology for Guam and neighboring islands. Because of the amount of the waste generated on Guam and the limited space available for landfilling, the Isfahan waste management technology appears to be the most practical and feasible method that Guam and other islands in Micronesian could possibly adopt as a sustainable waste-management strategy.

REFERENCES

1. United Nations Environmental Program (2009) Developing Integrated Solid Waste Management (ISWM) Plan Training Manual. Division of Technology, Industry and Economics, International Environmental Technology Centre Osaka, Shiga Japan 4:1-172
2. Bureau of Statistics and Plans (2010) Guam Census Population Counts.
3. Guam Solid Waste Authority (2014) Guam Solid Waste Authority Services.
4. Guam Solid Waste Receivership Information Center (2009).
5. Guam Integrated Solid Waste Management Plan (2006).
6. Johnson K (2009) Waste generation rate for Guam residents personal observation.
7. European Commission documentation (2004) Methodology for the Analysis of Solid Waste (SWA-Tool) User version. Project: SWA-Tool, Development of a Methodological Tool to Enhance the Precision & Comparability of Solid Waste Analysis Data. 57.
8. Martinho MGM, Silveira AI, BrancoFDEM (2008) Report: New guidelines for characterization of municipal Solid Waste: the Portuguese case. Waste Management and Research 265:486-490.
9. Isfahan Composting Factory. Steps taken during the zero waste management strategy as described by the general manger.

PART IV

PYROLYSIS AND CHEMICAL UPGRADING

CHAPTER 7

Furfurals as Chemical Platform for Biofuels Production

DANIEL E. RESASCO, SURAPAS SITTHISA, JIMMY FARIA, TEERAWIT PRASOMSRI, AND M. PILAR RUIZ

7.1 INTRODUCTION

The conversion of lignocellulosic biomass to liquid fuels has attracted renewed attention in recent years due to its environmental, economic, and strategic advantages [361–3]. In contrast to fossil fuels, the biomass-derived fuels can be CO_2–neutral, since the CO_2 produced in their combustion can be reabsorbed by green plants and algae during photosynthesis [4–7]. However, there are technical and economical challenges that have delayed the application of this technology to the commercial scale. An economically sustainable and competitive process will require versatility to accept different kinds of biomass feedstocks and the ability to produce low-oxygen, high energy content liquids, which should be fungible with conventional fuels [8].

Furfurals as Chemical Platform for Biofuels Production. Resasco DE, Sitthisa S, Faria J, Prasomsri T, and Pilar Ruiz M. In Heterogeneous Catalysis in Biomass to Chemicals and Fuels *(2011), Ed. David Kubička and Iva Kubičková, ISBN: 978-81-308-0462-0. Reprinted with permission from the authors.*

There are different possible sources of biomass that could be used as feedstock in the production of biofuels. They include waste materials (agricultural or urban), forest products (wood, logging residues, trees, shrubs), energy crops (starch crops such as corn, wheat, barley; sugar crops; grasses; woody crops; vegetable oils; hydrocarbon plants), or aquatic biomass (algae, water weed, water hyacinth) [8]. The availability of enough quantities of biomass is a key factor in scaling up biofuels production. In the U.S., it is estimated that 1.3×10^9 metric tons of dry biomass/year could be sustainably produced for biofuels, without a significant impact on human food, livestock feed, and export demands [9]. This amount of biomass would represent an energy content of 3.8×10^9 boe (barrels of oil energy equivalent), which would account for 54 % of the current annual demand of crude oil in the U.S., Ref. [10,11].

Gasoline and diesel are commodities that have been optimized through many years of commercial practice and dedicated research. The notion of molecular management and molecular engineering of fuels has been put into action in refining operations of petroleum fuels. These terms entail having the right molecule in the right place, at the right time, and at the right price [12], and at even higher level of molecular manipulation, they imply a purposeful design of molecules with precise structures and well-defined properties. To achieve this high level of chemical specificity, the continuous development of better catalytic materials is crucial [13]. The arrival of biofuels in the energy scene, particularly those from second- and third-generation technologies, produced from non-food biomass resources, poses new challenges and research opportunities for fuel development and catalytic upgrade [14]. Among the properties that are required to qualify a given fuel type, one can mention octane number (ON), cetane number (CN), sooting tendency, water solubility, freezing point, viscosity, flash point, cloud point, autoignition temperature, flammability limits, sulfur content, aromatic content, density, boiling temperature, vapor pressure, heat of vaporization, heating value, thermal and chemical stability, and storability. Many of these properties can be improved by catalytic upgrading. Therefore, when designing an upgrading strategy, a refiner is required to know how each of these properties will be affected by the structure of the components that are being allowed into a given fuel. Thus, it is important to develop predicting tools to determine how the structure of molecule added to the fuel will affect each of the properties of interest. Since for many fuel properties the overall value of the property for the mixture depends non-linearly on the individual contributions, the refiner also needs to be able to predict how new components behave in fuel mixtures [15].

The molecular engineering approach has been applied for upgrading of fossil fuels and there are many examples in the literature. However, the application of this rational approach in the upgrading of biofuels is just starting [16]. To optimize a fuel property of interest it is necessary to develop first an experimental database of known properties for components chemically similar to those to be incorporated in the fuels. Then, one needs to use correlations such as Quantitative Structure Property Relationships (QSPR) to expand the database to all the possible components of the fuel. QSPR utilizes molecular descriptors, (i.e., numerical values calculated from the molecular structure) to develop correlations with specific properties of the corresponding compounds. Molecular descriptors involve geometric, steric, and electronic aspects of the molecule and can range from very simple physical parameters such as the number of carbon atoms or branches in a molecule, to more complex parameters such as dipole moment or surface area. Application of QSPR to fuel properties has resulted in models that can estimate Cetane Number [17,18], Octane Number (RON and MON) [19], and sooting tendencies [20] of any fuel component, only on the basis of the molecular structure. With these structure-property relationships the researcher can evaluate whether a given potential catalyst and conversion process modify the structure of a given reactant in a favorable way regarding the desired fuel properties [21].

In this chapter we compare different conversion strategies involving furfural (FAL) and hydroxymethyl furfural (HMF), two chemical building blocks for the production of transportation fuels as well as a variety of useful acids, aldehydes, alcohols, and amines [22,23]. For example, C–O hydrogenolysis of HMF produces 2,5-dimethylfuran (DMF), a potentially good gasoline component, with heating value (35 MJ/Kg) and boiling point (93 °C) higher than those of ethanol (22.6 MJ/Kg and 78 °C, respectively) [24]. Being immiscible in water makes it even more suitable for transportation fuels. Similarly, the hydrodeoxygenation of the trimer obtained from the aldol-condensation of two molecules of furfural and one of acetone yields linear paraffins that could be used as jet or diesel fuel components, since they have the required properties for these fuels [25].

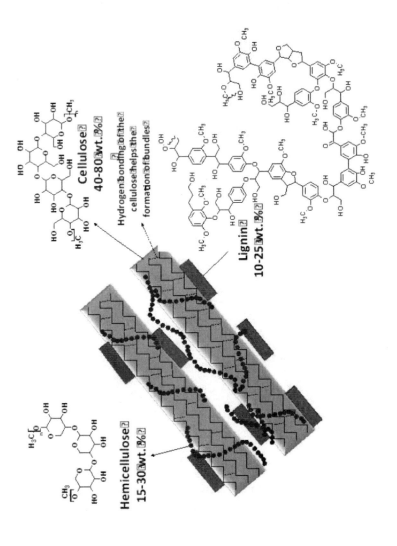

FIGURE 1: Schematic representation of the composition of lignocellulosic biomass.

In the following sections we will first describe how furfurals are obtained from biomass and then compare potential upgrading strategies to produce transportation fuel components with desirable properties.

7.2 BIOMASS TO FURFURALS

Photosynthesis uses solar radiation to convert energy-poor carbon dioxide and water into energy-rich carbohydrates $(CH_2O)_n$ and molecular oxygen (O_2), Ref. [26]. These carbohydrates are stored in the plants in the form of polysaccharides that include starch, cellulose, and hemicellulose and account for up to 75 % of the plant mass. Although the energy efficiency of the photosynthesis process varies from 0.1 to 8.0 %, the areal growth rate of the plants is usually very high (6 to 90 metric tons/ha-year).

Polysaccharides in plants have very distinct functions and their molecular structures are the most appropriate for each purpose. For example, to give structural resistance to the plant, cellulose is a large and well-organized polymer and is located in the primary cell wall. In contrast, hemicellulose is a branched polysaccharide that is less rigid and able to wrap around the cellulose structure (Figure 1). Finally, starch is a polymer of d-glucose with α-1,4- glycosidic bonds (repeating unit $C_{12}H_{16}O_5$), Ref. [8,27]. Those linkages occur in chains of α-1,4 linkages with branches formed as a result of α-1,6 linkages, making its structure highly amorphous and readily attacked by enzymes in the digestive system of animals and humans. Starch is a main constituent of corn, rice, potato, etc. Although it has been widely used for fuels and chemicals production [10,28], starch should not be replaced as a source of food [29].

Another major component of biomass is lignin (10–25 %), which has a noncarbohydrate polyphenolic structure that is encrusted in the walls of the cell and cement it together (Figure 1). This complex, cross-linked, highly aromatic structure (molecular weight ~10,000 g/mol) is derived from three monomers called monolignols (coniferyl alcohol, sinapyl alcohol and p-coumaryl alcohol). Coniferyl alcohol is the predominant lignin monomer found in softwoods, while p-coumaryl alcohol is a minor component of grass and forage type lignins. Both coniferyl and sinapyl alcohols are the building blocks of hardwood lignin [30,31]. Other compounds present in biomass in smaller quantities include pigments, sterols, triglycerides, terpenoids, resins, waxes, alkaloids, and terpenes.

7.2.1 PRIMARY CONVERSION OF BIOMASS

Although lignocellulose is one of the most abundant sources of biomass, its conversion into liquid fuels is a challenging task [32]. The more reactive sugar polymers in the plant are protected by the highly unreactive lignin fraction. For that reason, an effective pretreatment step is necessary to break down the refractory lignin seal, disrupt the crystalline structure of the cellulose, and increase the surface area of the biomass. A variety of methods are available to perform this pretreatment including immersion in hot water, dilute acid solutions, caustic lime/ammonia, and steam explosions [33,34].

Figure 2 summarizes the three major approaches suitable for primary processing of lignocellulosic biomass [35]. They are gasification to syngas [36,37], liquefaction (pyrolysis) to bio-oil [38] and hydrolysis for production of sugars [39]. Afterwards, further refining is necessary (secondary processing) to bring the mixture to the range of the desired fuel by reducing the oxygen concentration and maximizing the energy content of the liquid.

Several routes have been proposed for the secondary processing. They include (a) conversion of biomass-derived syngas into alkanes or methanol by Fischer-Tropsch or methanol synthesis [40-41], (b) catalytic aromatization, condensation and hydrodeoxygenation of bio-oil for hydrocarbons production [42–48], (c) enzymatic fermentation of sugars for methanol and butanol generation [49], (d) hydrogen synthesis by aqueous phase reforming of sugar-derived species [50], (e) conversion of monosaccharide molecules into aromatics and coke by zeolite upgrading [51–53], (f) fuels production by biphasic processing of sugars and oxygencontaining molecules [54–55], and (g) esterification of triglycerides for biodiesel synthesis [56].

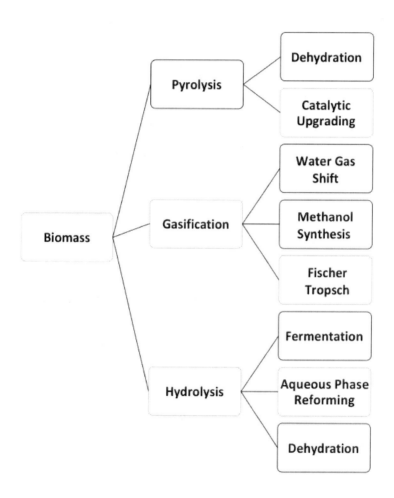

FIGURE 2: Different strategies for primary conversion of lignocellulosic biomass.

Gasification is outside the scope of this chapter and the reader is referred to excellent reviews on this subject [37,57,58]. We will briefly discuss the other two primary conversion processes, with emphasis on those that lead to the production of furfurals.

7.2.1.1 PYROLYSIS OF LIGNOCELLULOSIC BIOMASS

Pyrolysis oil is produced during the fast heating of lignocellulosic biomass in the absence of oxygen. Depending on the feedstock and the pyrolysis conditions, mainly temperature, heating rate, and residence time, it is possible to produce yields of bio-oil ranging from 50 to 95 wt. % [59]. Biooil is a dark brown, highly viscous, and oxygenated liquid with up to 40 wt. % water content [60,61]. In addition to water, pyrolysis oil comprises catecols, syringols, guaiacols, phenolics, sugars, ketones, aldehydes, short carboxylic acids, and furfurals [62]. This mixture of highly reactive molecules makes the bio-oil an unstable system that requires further processing before its utilization for transportation fuels [38,63].

Pyrolysis of pure cellulose yields levoglucosan (up to 60 wt. %), via intramolecular condensation and sequential depolymerization of glycosidic units [64]. However, in the presence of alkali metals in concentrations as low as 0.005 mmoles/g, non-selective homolytic cleavage of carbon-carbon bonds occurs, favoring the formation of low molecular weight molecules including acids, furfurals, and linear aldehydes (Figure 3) Refs. [65–69].

These undesired reactions are controlled by the ionic nature of the inorganic salts, the Lewis acidity/basicity and/or their ability to form complexes that stabilize particular reaction intermediates. For example, when $CaCl_2$, $Ca(NO_3)_2$ and $CaHPO_4$ were added during the pyrolysis of cellulose, the yield of glycolaldehyde was reduced, while a remarkable increase was observed in the yield of formic acid and acetol. More interesting, the formation of furfural, hydroxymethyl furfural, and levoglucosenone is enhanced in the presence of $CaCl_2$, which indicates the participation of these salts in the dehydration reaction of the cellulose [67].

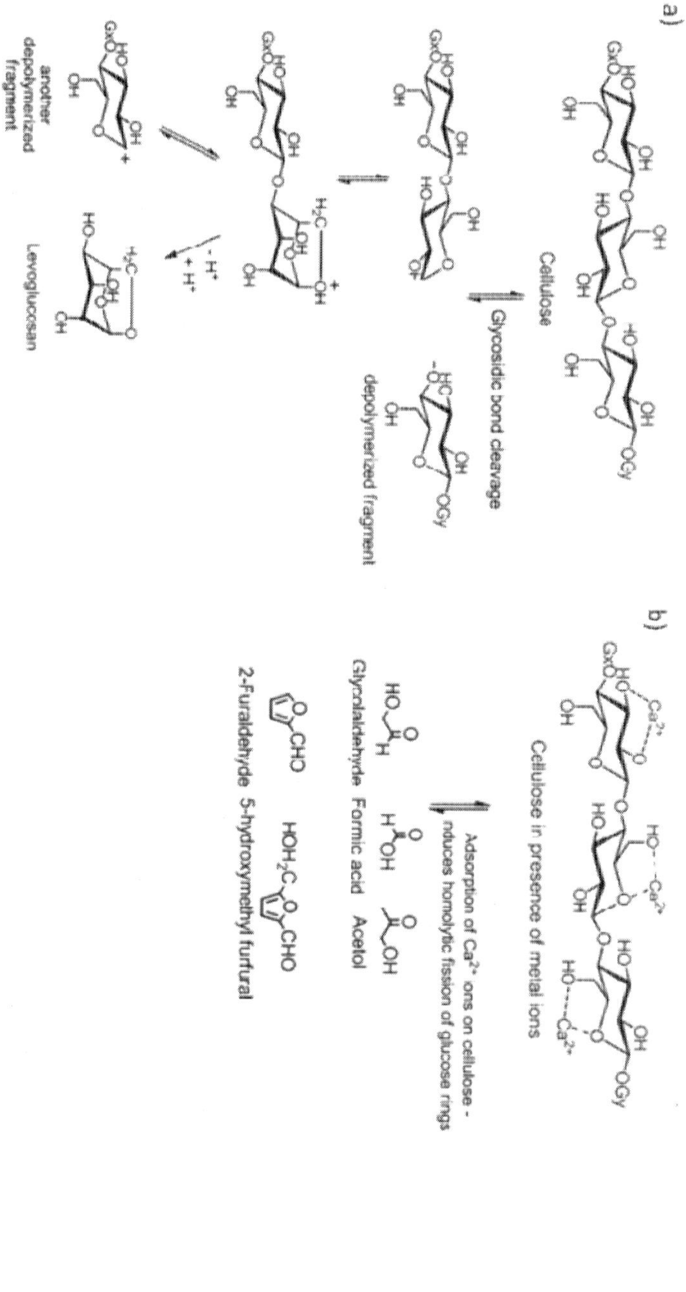

FIGURE 3: Reaction scheme for cellulose conversion a) in the absence and b) in the presence of metal ions (adapted from Refs. [66,68,69]).

7.2.1.2 HYDROLYSIS

The deconstruction of lignocellulosic biomass via hydrolysis is readily catalyzed by strong acids, which speed up the protonation of the oxygen bridges connecting the sugar monomers. Mineral acids (HCl, H_2SO_4, and HNO_3) are effective catalysts for depolymerization of hemicellulose and cellulose, as they have the necessary strength to break down the crystalline structure of the cellulose fibers. Acid catalyzed hydrolysis has been conducted in industrial scale for a long time. In 1922, The Quaker Oat Company developed the first commercial process for production of furfural from oat hulls in Cedar Rapids, Iowa, using liquid acid catalysts [70]. The first production of furfural was achieved by passing superheated steam through biomass. However, at high enough temperatures, sulfuric acid is not necessary since the acetic acid released during the depolymerization of hemicellulose is able to catalyze the dehydration reaction and produce furfural.

The acid catalyzed hydrolysis of polysaccharides comprises the following steps [71], (1) protonation of an oxygen link leading to a trivalent oxygen, (2) heterolytic cleavage of a C–O bond to form a carbocation and a hydroxyl group, (3) reaction of a water molecule with the carbocation, and (4) release of a proton from the resulting H_2O+ with consequent regeneration of the catalyst and yield of a hydroxyl group. The sequence is repeated until the polysaccharide is fully converted into the corresponding sugars. However, during this process the lignin fraction cannot be dissolved and remains unreacted. Therefore, further refining is necessary. The catalytic effect of the acid is highly dependent on the proton transfer efficiency of the catalyst, rather than the proton concentration in the system. In fact, it has been observed that even negligible hydrogen ion concentrations may have a catalytic effect [72]. In this context, the hydrogen ion concentration is highly dependent on the temperature. It has been observed that the dielectric constant of water, responsible for the dissociation of the acids, decreases with temperature, which strongly diminishes the dissociation of the acid [73].

Furfurals as Chemical Platform for Biofuels Production 113

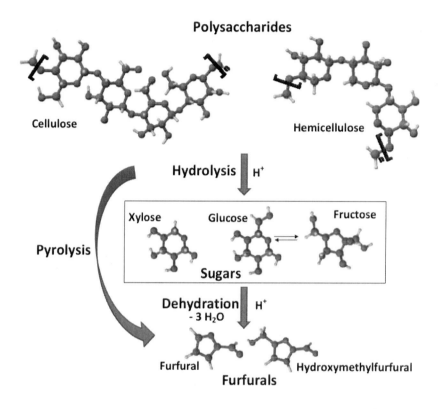

FIGURE 4: Production of furfurals (2-furfuraldehyde and hydroxymethylfurfural) from lignocellulosic biomass.

7.2.2 DEHYDRATION OF SUGARS

The formation of furfural and hydroxymethylfurfural can be accomplished by triple dehydration of xylose (pentose) and glucose (hexose), respectively, in the presence of an acid catalyst (Figure 4). Pentoses and hexoses are mostly present in the ring form, since the open-chain aldehydes only amount to less than 1 wt. %, Ref. [74]. Isomerization of glucose to fructose is desirable because the latter is more effective for production of HMF. For example, Sn-containing zeolites have been found to be very active and selective towards fructose (31 wt. %), with small production of mannose (9 wt. %) at mild temperatures (110–140 °C), [75].

The transformation of the C_5, C_6 sugars is performed through the sequence of two 1,2-eliminations, followed by 1,4-elimination of water molecules. In the final step, ring formation occurs facilitated by the tendency of sp^2 carbon atoms to form planar structures [76]. After the 1,4-elimination, the hydrogen ion is eliminated and furfurals are obtained.

Ionic liquids have also been tested for the production of furfurals [77–79]. In this approach, the ionic liquid stabilizes the furfural (FAL) or hydroxymethylfurfural (HMF) product from the reaction mixture and increases the reaction selectivity. However, the detailed role of the ionic liquid in the catalytic cycle is still under study [80].

Possible mechanisms have been proposed for the dehydration of fructose in the presence of halides, as well as for the isomerization of glucose to fructose catalyzed by hexacoordinated chromium (II) complexes (Figure 5), Ref. [79]. It appears that the halide (X-) attacks the oxocarbenium ion to form a 2-deoxy-2-halo-intermediate that is more stable and, therefore, less prone to oligomerization and reversion reactions (Figure 5a).

An alternative explanation has been given by Zhao et al. [77] (Figure 5b), who proposed that, when ionic liquids like 1-methyl-3- methylimidazolium chloride [EMIM]Cl are mixed with chromium chloride ($CrCl_2$), the Lewis acid formed ($CrCl_3^-$) plays a critical role in the proton transfer, which facilitates the mutarotation of glucose.

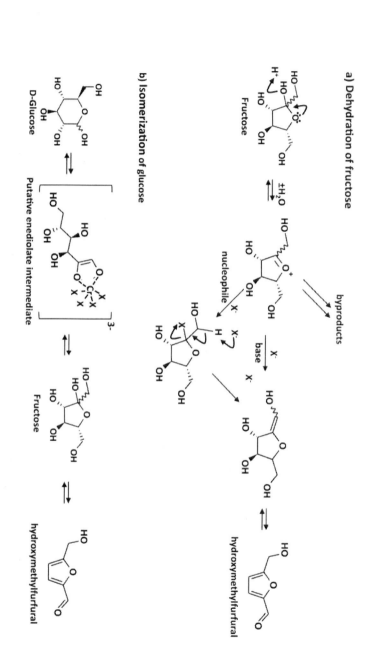

FIGURE 5: Putative mechanisms of dehydration of fructose and isomerization of glucose in the presence of halides and hexacoordinate chromium (II) complex, respectively. [77, 79].

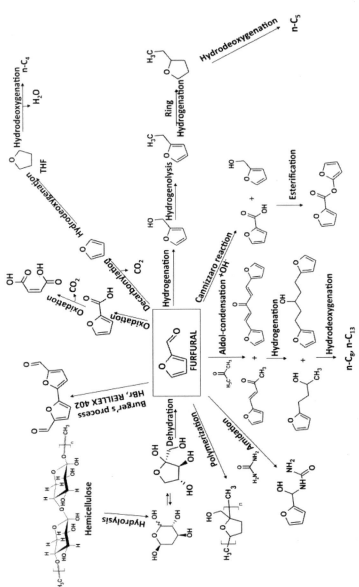

FIGURE 6: Catalytic pathways for the conversion of furfural.

7.3 CATALYTIC STRATEGIES FOR CONVERSION OF FURFURALS TO FUEL COMPONENTS

Several pathways for conversion of furfurals are considered in Figures. 6 and 7. Among these routes, aldol-condensation has been proposed to increase the molecule length through the formation of carbon–carbon bonds [81,82], and etherification has been presented as an alternative to produce specialty chemicals for pharmaceutical industry [83].

7.3.1 ALDOL-CONDENSATION

Biomass-derived sugars can be upgraded to conventional fuels through the combination of aldol-condensation reactions and hydrogenation/dehydration reactions. Dumesic et al. [35,54,84–87] have done seminal work in this area, applying molecular engineering concepts to formulate refining strategies for the production of renewable fuels.

The aldol-condensation reaction is initiated by the abstraction of the α-hydrogen from aldehydes and ketones, in the presence of a basic catalyst. The resulting enolate reacts with the carbonyl carbon of another molecule to form the aldol-product that, after dehydration, forms an α,β-unsaturated carbonyl molecule.

Although furfural molecules do not have α-hydrogens, it is possible to perform cross-condensation reactions with other aldehydes and ketones, like acetone and propanal. In this approach, diesel and jet fuel can be produced by the coupling of C_5–C_6 furfural molecules with C_3 aldehydes and ketones (Figs. 6 and 7, respectively), Ref. [88]. The cross-condensed products, ranging from C_8 to C_{15}, can be hydrogenated at low temperatures (100–150 °C) to decrease their solubility in the aqueous phase and their oxygen content. High temperatures and bifunctional catalysts (acid and metal) are required to perform dehydration and hydrogenation reactions in order to remove the residual oxygen and produce C_8 to C_{15} alkanes. Another route proposed by these authors [88] was based on the partial hydrogenation of the aromatic ring in the furfural molecules, in which an α-hydrogen is generated and 5-hydroxymethyl-tetrahydrofurfural

(HMTHFA) and tetrahydro-2-furfural (THF2A) are formed as products. These molecules can then react by self aldol-condensation to generate C_{10} to C_{12} products.

The aldol-condensation reactions are carried out in polar solvents, such as water or water-methanol, and usually catalyzed by homogeneous bases like NaOH or $Ca(OH)_2$, Ref. [89]. However, in homogeneous catalysis, for every 10 parts of product formed, one part of spent catalyst is generated, which increases up to 13 % the selling price of the product due to required purification, recovery, and waste treatments [89]. In contrast, heterogeneous catalysts are readily separated by filtration methods and, depending on the reactor configuration, they can be used for vapor-phase [90,91], liquid-phase [92,93] or biphasic liquid [55,82] systems.

7.3.2 ETHERIFICATION

The production of chemical intermediates derived from fossil resources is of great importance for the farmaceutical and polymer industries, as more than 90 % of the major organic chemicals manufacturated in the US in 1995 were produced from conventional hydrocarbon sources [94]. Nevertheless, the diminishing of the crude oil reserves and the ever-increasing environmental regulations has boosted the interest in the conversion of biomass into chemicals [95]. A perfect example of this is the etherification of hydroxymethylfurfural to produce 5,5-oxy(bis-meth-ylene)-2-furaldehyde (OBMF), which can be used in the synthesis of imine-based polymers with high glass transition temperatures (300 °C), as well as in the preparation of hepatitis antiviral precursors [83]. In that case, the catalysts used were zeolites and mesoporous aluminosilicates with Brønsted and Lewis acid sites, which offer high temperature resistance and size-shape selectivity [96].

Two different etherification processes can be found in the literature (Figure 8), Ref. [83]: 1) etherification of HMF catalyzed by acids in organic solvents [97–99] and 2) the Williamson reaction of HMF and 5-chloromethyl- 2-furfural in an excess of base [100].

Furfurals as Chemical Platform for Biofuels Production

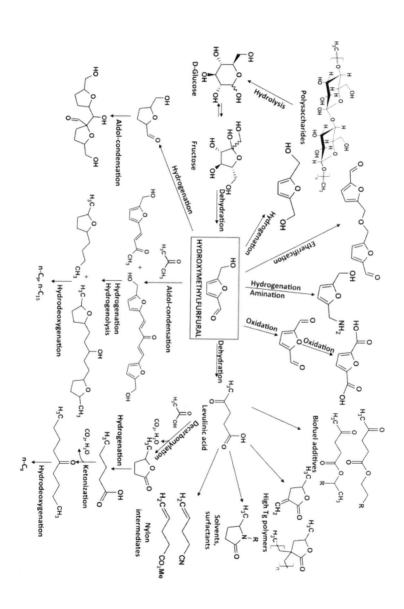

FIGURE 7: Catalytic pathways for the conversion of hydroxymethylfurfural.

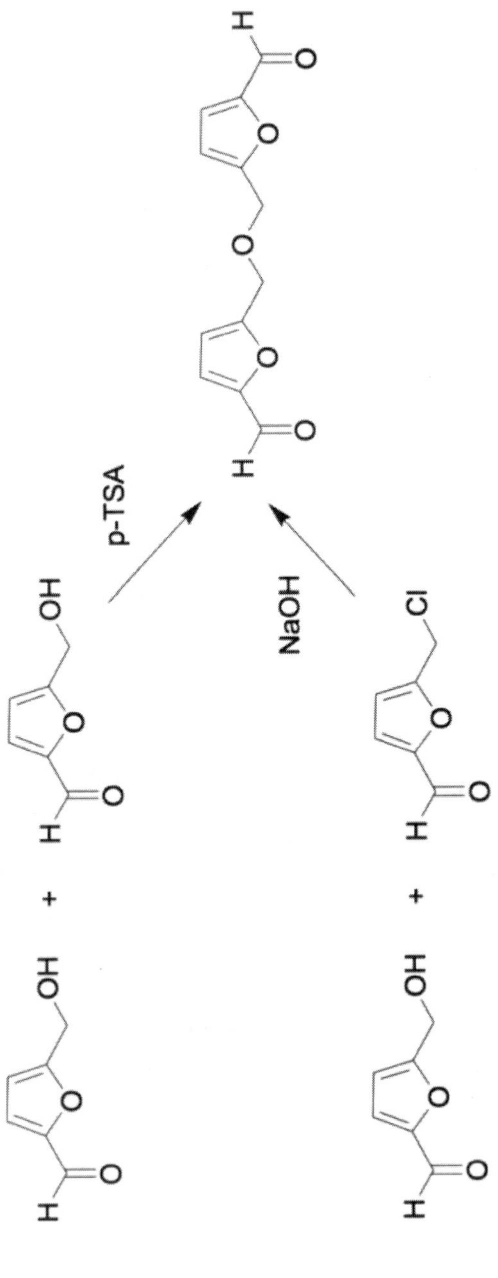

FIGURE 8: Synthetic pathways for the production of 5,5'-oxy(bis-methylene)-2-furaldehyde (OBMF). Reproduced from Ref. [83].

7.3.3 HYDROGENATION

Hydrogenation of furfural with hydrogen gas on metal catalysts produces mainly furfuryl alcohol. This alcohol finds uses as a solvent, but is primarily used as an ingredient in the manufacture of chemical products such as foundry resins, adhesives, and wetting agents [74]. The catalysts most typically used in furfural hydrogenation are metals because they are able to dissociate hydrogen, thus making hydrogenation possible. The choice of catalyst support is based on its ability to disperse and stabilize metal particles enhancing the active surface area. Compared to other aldehydes, in addition to the carbonyl group, furfural contains an aromatic furanyl ring that could b also hydrogenated. While carbonyl hydrogenation is usually preferred due to the high stability of the aromatic ring, metal catalysts that have strong interactions with the unsaturated C=C bonds can still saturate the ring. Therefore, the selectivity toward aromatic alcohols is strongly dependent on the metal catalyst used. Furthermore, the geometric and electronic properties of different metals can affect both hydrogenation activity and selectivity by influencing the type of adsorption intermediates. In this section we will primarily discuss the differences observed in furfural hydrogenation on group IB and VIII metals.

Among the group IB metals, Cu has been the most intensively investigated as a catalyst for furfural hydrogenation [101–107]. Silver has been used in a few studies while there is no report on Au catalysts [108]. The group IB metals are significantly less active than other metals. However, they exhibit a remarkable selectivity towards hydrogenation of the carbonyl group leaving the C=C double bonds in the furanyl ring unreacted. In this sense, Cu has been found to be the most selective among all tested metal catalysts. Selectivities above 98 % to furfuryl alcohol have been achieved over monometallic Cu/SiO_2 catalysts [107]. Also, monometallic Ag catalysts have been found able to hydrogenate the C=O group of furfural with relatively good selectivity, but not as high as that of Cu. For example, 80 % selectivity was obtained over a Ag/SiO_2 catalyst prepared by the sol-gel method [108].

The group VIII metals (Ni, Pd, and Pt) have also been used for furfural hydrogenation in both vapor and liquid phases on different supports;

they all exhibit significant activity and selectivity [108], particularly at low temperatures (i.e., below 200 °C). For example, Ni catalysts have been used at low temperatures (100 °C) in liquid phase, exhibiting > 95 % selectivity to furfuryl alcohol [109,110]. Likewise, supported Pt catalysts doped with transition metal oxides have exhibited high selectivity, e.g., Pt/TiO$_2$/SiO$_2$ (selectivity = 94 %) and Pt/ZrO$_2$/TiO$_2$ (selectivity = 95 %), Ref. [111]. However, at higher temperatures (> 200 °C) the selectivity to alcohols on group VIII metals drops significantly upon the appearance of other reactions. For example, over a Ni/SiO$_2$ catalyst at 230 °C and excess of hydrogen (H$_2$/feed ratio = 25), only 25 % selectivity to furfuryl alcohol was observed [112]. On Pd catalysts, it is even lower, e.g., a Pd/SiO$_2$ catalyst at 230 °C yielded 14 % selectivity [113] and a Pd-Y catalysts at 350 °C rendered only 1 % selectivity to furfuryl alcohol [101]. The lower selectivities observed at high temperatures on group VIII metals can be ascribed to the emergence of decarbonylation and ring opening reactions. In addition, hydrogenation of the furanyl ring yielding the saturated alcohol is also observed over group VIII metal catalysts due to a stronger interaction of the furanyl ring with the metal surface than that obtained with group IB metals [114]. As a result, Cu and Ag should be the preferred catalysts if high selectivity to furfuryl alcohol is desirable, particularly at high temperatures. However, the intrinsic activity for furfural hydrogenation is lower on group IB metals than on group VIII metals, which is typically explained by the difference in extent of d-orbital filling. That is, the d-orbitals are filled in group IB, reducing the bond strength.

7.3.3.1 REACTION INTERMEDIATES AND MECHANISMS FOR FURFURAL HYDROGENATION

DFT (density functional theory) calculations and HREELS (High Resolution Electron Energy Loss Spectroscopy) studies have been used to investigate the reaction intermediates and mechanisms for the hydrogenation of furfural on metal surfaces. For example, the preferential adsorption configurations of the furfural molecule on a Cu(111) surface have been investigated by DFT [107]. These calculations have shown that the perpendicular adsorption mode (see Figure 9) is the most stable configuration with the carbonyl group directly bonded to the surface through the O lone pair, which acts as a Lewis base. This type of adsorption mode is termed η1-(O) and it seems to occur exclusively on group IB metals [115].

In contrast, the parallel ring adsorption is highly unstable on the Cu(111) surface. This instability arises from the overlap of the filled 3d band of the surface Cu atoms with the π orbitals of the aromatic furanyl ring, which causes a repulsion that results in a net positive (i.e., endothermic) adsorption energy.

A simple analysis of these results leads to a clear explanation of the observed high hydrogenation selectivity of the carbonyl group on these metals. That is, since the most stable configuration over group IB metals is the η1-(O) mode, only the carbonyl group can be hydrogenated, while the predicted repulsion between the aromatic ring and the Cu surface makes the hydrogenation of the furanyl ring much more difficult.

Starting from the η^1-(O) adsorbed species on Cu surface two different reaction paths have been calculated by evaluating the possible transition states [107]. As shown in Figure 10, the hydrogen attack to the carbonyl group can occur in two different ways. In the first case (mechanism (a)), the addition of the H atom first occurs on the C atom of the carbonyl, leading to an alkoxide. The second H atom is added to the O of the alkoxy intermediate, yielding the alcohol product. In the second case (mechanism (b)), hydrogenation occurs first on the carbonyl O atom with formation of a hydroxylalkyl intermediate, followed by addition of the second H atom to the carbon to produce the furfuryl alcohol.

The DFT calculations indicate that the first H attack to the O atom of the carbonyl (hydroxyalkyl species, mechanism (b)) would have a lower energy barrier than the first H attack to the C atom of the carbonyl (alkoxy intermediate, mechanism (a)). The difference in stability between the two intermediates can be ascribed to the role played by the aromatic furanyl ring. That is, the addition of the first H atom to the O atom produces a negative charge on the C atom, which can be stabilized by the presence of furanyl ring (delocalization), favoring the formation of the hydroxylalkyl intermediate. This stabilization would not occur with aliphatic aldehydes, for which only mechanism (a) is possible.

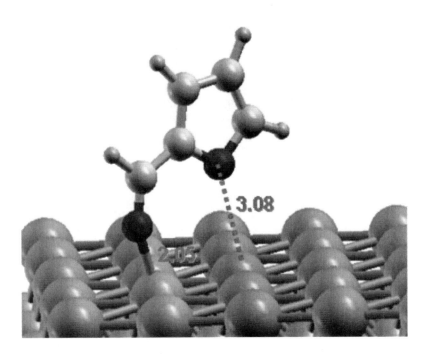

FIGURE 9: Adsorption geometry of furfural on Cu(111), Ref. [107].

As opposed to group IB metals, on which η^1-(O) aldehyde is the preferred surface species, clean group VIII metal surfaces tend to adsorb aldehydes in the so-called η^2(C,O) configuration, in which both C and O atoms of the carbonyl group are bonded to the metal surface. In fact, DFT calculations of furfural on Pd(111) clearly show that an η^2(C,O)-aldehyde is the preferred configuration on Pd (Figure 11), Ref. [113]. These results fully agree with the experimental HREEL spectroscopic observations of Barteau et al. [116]. The DFT calculations also demonstrate that the preferred adsorption of furfural is with the furanyl ring oriented essentially parallel to the metal surface. Therefore, this configuration makes hydrogenation of the furanyl ring on Pd readily possible, yielding the saturated alcohol as a major product, as experimentally observed.

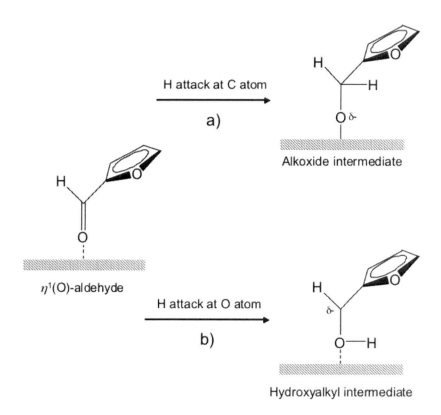

FIGURE 10: Formation of a) alkoxide and b) hydroxylalkyl intermediates over Cu/SiO$_2$ [107].

Th first H attack to the O atom of the η^2(C,O) aldehyde can result in a C-bonded hydroxyalkyl surface species, which can become the reaction intermediate in the hydrogenation of furfural on Pd to produce furfuryl alcohol. However, when the temperature is high enough, the η^2(C,O) aldehyde may further convert on the surface to a more stable acyl species, which is a precursor for decarbonylation, leading to hydrocarbon fragments and CO [117]. This tendency lowers the alcohol selectivity on Ni and Pd catalysts, at the expense of furan, which is the direct product of decarbonylation, discussed in Section 3.4.

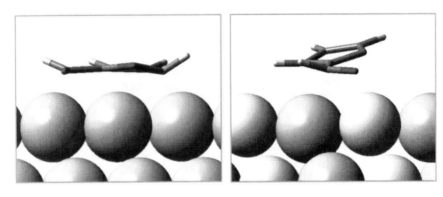

E_{ads} = 55.4 kcal/mol E_{ads} = 9.6 kcal/mol

FIGURE 11: Adsorption geometry of furfural on Pd(111) and PdCu(111), Ref. [113].

7.3.3.2 HYDROGENATION KINETICS

The heterogeneously catalyzed hydrogenation of furfural over supported surfaces exhibits several elementary steps, including adsorption of the reactants, surface reaction, and desorption of the products.

A Langmuir-Hinshelwood model is usually a good approximation to describe the kinetics of this reaction [118]. The model can contemplate competitive and non-competitive adsorption, as well as dissociative and nondissociative adsorption of reactants. In cases in which hydrogen is used in high excess, its partial pressure remains essentially unchanged and it is typically incorporated in the reaction constants. Furthermore, the kinetic model can assume one or two types of adsorption sites.

For example, a single-site Langmuir-Hinshelwood kinetic model based on dissociative adsorption of hydrogen was used by Sitthisa et al. [107] to describe the gas phase hydrogenation of furfural over Cu/SiO_2. In that study, the adsorption of hydrogen, the aldehyde, and the products was assumed to involve one type of site.

Adsorption of the organic compounds was assumed to be rapid compared to the reaction steps, implying quasi-equilibrium for the adsorption and desorption steps. That is, the irreversible hydrogenation step was assumed to determine the rate of product formation. In addition to furfuryl alcohol (FOL), methyl furfural (MF) is also obtained in small amounts by cleavage of the C–O bond over Cu/SiO_2. After writing the mass balance for consumption and formation of each compound (FAL, FOL, and MF), the following rate equations were obtained:

$$r_{FAL} = \frac{-k_1 K_{FAL} P_{FAL} + \frac{k_1}{K} K_{FOL} P_{FOL}}{1 + K_{FAL} P_{FAL} + K_{FOL} P_{FOL} + K_{MF} P_{MF} + K_{H_2}^{1/2} P_{H_2}^{1/2}} \quad (1)$$

$$r_{FOL} = \frac{k_1 K_{FAL} P_{FAL} - \left[\frac{k_1}{K} K_{FOL} P_{FOL} + k_2 K_{FOL} P_{FOL}\right]}{1 + K_{FAL} P_{FAL} + K_{FOL} P_{FOL} + K_{MF} P_{MF} + K_{H_2}^{1/2} P_{H_2}^{1/2}} \quad (2)$$

$$r_{MF} = \frac{k_2 K_{FOL} P_{FOL}}{1 + K_{FAL} P_{FAL} + K_{FOL} P_{FOL} + K_{MF} P_{MF} + K_{H_2}^{1/2} P_{H_2}^{1/2}} \quad (3)$$

where k_1 and k_2 are the rate constants for hydrogenation and hydrogenolysis, respectively (including hydrogen pressure), K_i is the adsorption constant and P is the partial pressure of component i (FAL = furfural, FOL = furfuryl alcohol, MF = 2-methyl furan).

By fitting the kinetic data with a LH model, all the thermodynamic and kinetic parameters were obtained. They include the heat of reaction (ΔHreaction = -11.9 kcal/mol), the heats of adsorption of furfural (12.3 kcal/mol) and 2-methyl furan (3.7 kcal/mol), Ref. [107]. The calculated heat of reaction was in the same order of that experimentally measured in the production of furfuryl alcohol over a copper chromite catalyst (-14.5

kcal/mol), Ref. [119]. The significant difference between the heats of adsorption of furfural and 2-methyl furan can be explained from the DFT calculations, which as indicated above, show that the strongest interaction of furfural with the Cu surface is via the carbonyl O forming η^1-(O) aldehyde. Since methyl furan does not have a carbonyl O and the furanyl ring is repelled by the Cu surface, a very weak adsorption is expected. This behavior of Cu strongly contrasts with that of other group VIII metals such as Pd or Pt, which strongly bind the furanyl ring [114].

Likewise, a similar Langmuir-Hinshelwood model of a single site has also been applied for the hydrogenation of furfural towards furfuryl alcohol over Ir/TiO$_2$ [120]. In this study, the adsorption of furfuryl alcohol was not taken into account in this kinetic model and the rate expression was expressed as:

$$r_{FAL} = \frac{k_1 K_{FAL} C_{FAL} K_{H_2}^{1/2} P_{H_2}^{1/2}}{1 + K_{FAL} C_{FAL} + K_{H_2}^{1/2} P_{H_2}^{1/2}} \tag{4}$$

Two alternative single-site mechanisms (with and without dissociation of hydrogen) were proposed in a furfural hydrogenation study conducted in the liquid phase over a Pt/C catalyst [121]. From these models, rate expressions were developed for cases in which either the adsorption of furfural, or hydrogen, or the surface reaction was rate-controlling. However, all these cases resulted in some negative parameters and hence they were rejected. By contrast, a model involving a dual-site mechanism with hydrogen molecularly adsorbed on active sites different from those involved in the adsorption of furfural and furfuryl alcohol resulted in all positive parameters. In this case, the surface reaction between adsorbed furfural and adsorbed hydrogen was assumed to be rate-controlling. The rate expression derived for this case is

$$r_{FAL} = \frac{k_1 K_{FAL} K_{H_2} C_{FAL} C_{H_2}}{(1 + K_{FAL} C_{FAL})(1 + K_{H_2} C_{H_2})} \tag{5}$$

The activation energies (Ea) for hydrogenation of furfural to furfuryl alcohol obtained over different catalysts are summarized in Table 1.

TABLE 1: Activation energy of furfural hydrogenation for various catalysts.

Catalyst	Ea (kcal/mol)	ref.
Cu/SiO_2	11.8	107
Copper-chromite	11 ± 2	119
Ir/TiO_2	11.8	120
Ir/Nb_3O_5	~9	122
Pt/C	6.7	121

The activation energy for furfural hydrogenation on both Cu/SiO_2 and Cu-chromite catalysts is close to 12 kcal/mol. This activation energy is in the range of the values typically observed in the hydrogenation of other carbonyl compounds over Cu [119]. The significantly lower activation energy observed on Pt/C compared to Cu catalysts might be ascribed to the different adsorption phenomena occurring on Pt and Cu. Since on Pt the furfural adsorption occurs exclusively in the η^2 mode while on Cu it occurs in the η^1 mode, the carbonyl C and O atoms are closer to the metal surface in the former than in the latter. Therefore, it could be expected that the H attack is energetically unfavorable on Cu, which should result in a higher activation energy. The activation energies observed on Ir catalysts seem to depend strongly on the type of support used, i.e. Ir on TiO_2 and Nb_2O_5 supports [120,122]. That is, not only the type of metal but also the nature of the support may have a crucial effect on the hydrogenation reaction.

7.3.4 DECARBONYLATION

The decarbonylation of furfural results in the production of furan, which could be utilized as a fuel component, but also as a chemical intermediate. For example, furan can be subsequently hydrogenated to tetrahydrofuran, which is commonly used as solvent and as a starting material for polyurethane manufacture [74].

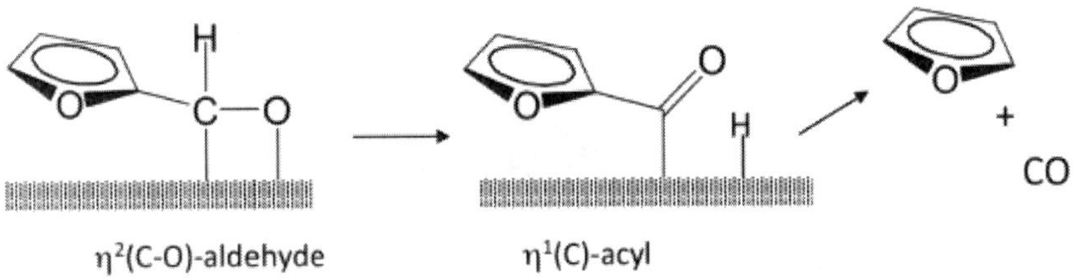

FIGURE 12: Proposed mechanism for decarbonylation of furfural over Pd/SiO$_2$ [113].

While decarbonylation of furfural can be carried out in both the vapor and liquid phases [123–125], the former is generally preferred due to the advantage of a simpler operation and the possibility of easier catalyst regeneration and reuse. Different catalysts have been tested for this reaction. They include metal oxides such as Mn, Zn, Cd, Sr, K, Fe, Mo, and Cr oxides [126–128] or group VIII metals such as Pd, Pt, Rh, Ni [112,123,129]. The latter are known to be effective catalysts and to result in high yields at rather mild conditions. By contrast, the former require more severe operating conditions and promote formation of heavy products that may cause catalyst deactivation.

Among the group VIII metals, Pd has been the most intensively investigated as the active metal in furfural decarbonylation [124,125] while Pt and Ni have been used less often [112,129]. It has been found that the high decarbonylation activity of Pd may be modified by the incorporation of a second metal (alloy formation) Ref. [113] or additives such as alkali and alkaline-earth metals [125,130]. Although Pd catalysts are relatively more active than other catalysts, the yield of furan decreases sharply with time on stream. For example, after 3 h on stream a Pd/C catalyst was found to lose most of its decarbonylation activity. An even faster deactivation was observed on a Pd/Al$_2$O$_3$ catalyst [123,127]. Among the various possible reasons for the observed deactivation, the most commonly proposed is carbon deposition, which could be due to side reactions, such as condensation and/or decomposition of furfural. Earlier studies demonstrated that a rapid loss in the activity of Pd/Al$_2$O$_3$ was observed in the absence of hydrogen, but the activity could be partially recovered by increasing the hydrogen partial pressure [124].

7.3.4.1 REACTION INTERMEDIATES AND MECHANISMS FOR FURFURAL DECARBONYLATION

The possible reaction paths for furfural conversion on Pd are illustrated in Figure 12. As mentioned above, the interaction of furfural with Pd involves a side-on complex (η^2-(C-O) aldehyde) in which the furan ring lies parallel to the surface. In this configuration, the carbonyl binds to the metal surface through the carbonyl π orbital, with overlap between d electrons of the metal and the π^* orbital of the carbonyl. The back-donation from the metal results in a stronger metal-aldehyde bonding than in the case of the η^1- mode. This stronger adsorption leads to a higher reaction rate, not only for hydrogenation of the η^2-(C-O) surface aldehyde, but also to its decomposition into a more stable acyl surface species (see Figure 12), in which the C atom of the carbonyl remains strongly attached to the surface. This acyl species may in fact be a precursor for the decarbonylation reaction, producing furan and CO. In fact, it has been shown that the selectivity to decarbonylation significantly increases as a function of temperature, which is consistent with an activated conversion of η^2(C,O) into the acyl species that increases decarbonylation while reducing hydrogenation [113].

The decarbonylation activity and selectivity of Pd catalysts may be modified by formation of bimetallic alloys and addition of additives such as alkali and alkaline-earth metals. For example, a comparison of the behavior of Pd-Cu alloys to that of pure Pd indicates that Pd may be electronically modified by the addition of Cu [113]. It has been observed that the stability of the di-sigma η^2-(C-O) species is greatly reduced on the Pd-Cu alloy due to a lower extent of back-donation. Theoretical calculation of the adsorption of furfural on Pd(111) and PdCu(111) surfaces suggest that the presence of Cu significantly reduces the interaction strength of furfural with the metal surface (see Figure 11). In fact, the heat of adsorption on the bimetallic surface is significantly reduced relative to that on the pure Pd. That is, the adsorption strength of furfural was predicted to drop from 13.2 kcal/mol on Pd(111) to 2.3 kcal/mol on PdCu(111). This dramatic reduction in the strength of interaction can be ascribed to the role played by the furanyl ring, which has a strong affinity for Pd, but not for Cu. As shown on the side view of Figure 11, the repulsion between the ring and the Cu atoms makes the molecule bends away from the surface.

Lopez and Norskov [131] have shown that alloying Pd with Cu causes a change in the position of the d-band center of Pd. This shift causes a lower extent of π-backdonation from Pd to π* orbital of the C=C and C=O bonds in the furfural molecule, which results in a significant weakening of the furfural adsorption strength. Moreover, relatively long distances are predicted by DFT between the carbonyl C and the metal surface in the case of the alloy, which greatly hinders the formation of the acyl intermediate, needed for decarbonylation. The destabilization of this crucial intermediate results in a drastic decrease in the rate of furan decarbonylation as the percent of Cu in the alloy increases.

A strong electronic modification of the properties of the Pd surface can also be obtained by doping it with K [130]. In this case, the Pd-K interaction results in an enhanced strength of furfural adsorption. This enhancement has been experimentally verified by the temperature-programmed surface reaction (TPSR) and FTIR. At the same time, animprovement in decarbonylation rate is observed in the presence of K. It has been proposed that this improvement is due to an enhanced stability of the acyl intermediate.

7.3.4.2 DECARBONYLATION KINETICS

Obtaining reliable kinetic parameters for furfural decarbonylation has been found to be more difficult than for furfural hydrogenation because the extent of catalyst deactivation has been much greater. In fact, while almost no deactivation was observed under conditions of hydrogenation reactions, almost 85 % of the initial activity was lost after 4 h under decarbonylation reaction conditions [123]. Two possible reasons have been suggested for this rapid deactivation. They are the formation of coke, which is more significant at the temperature needed for decarbonylation, and catalyst fouling by either adsorbed furfural or some product formed during the reaction, including some of the high-molecular-weight byproducts resulting from dimerization of furan [124,125].

A possible solution to avoid this problem has been attempted in a study of vapor phase decarbonyltion of furfural on Pd/Al$_2$O$_3$ catalyst [124]. These authors obtained rates of reaction at zero time on stream $(-r_{FAL})_0$ by extrapolating the curves $(X_{FAL})_0$ vs. W/F_{FAL0} using a second-order polynomial, followed by analytical differentiation and analyzing the data according to the method proposed by Froment and Hosten [132]. Using a Langmuir-Hinshelwood approximation they were able to satisfactorily fit the data on the basis of single-site mechanism that resulted in the expression:

$$(r_{FAL})_0 = \frac{kK_{FAL}P_{Furan}}{(1 + K_{FAL}P_{Furan} + K_{RS}P_{Furan})^2} \tag{6}$$

where K_{RS} is the adsorption constant of furan and CO. With this methodology, the authors were also able to develop kinetics and mechanism for the deactivation process. It was concluded that the deactivation occurred in parallel with the main reaction and the deactivation kinetics was governed by the reaction of two adjacent furfural molecules on the surface, resulting in the formation of the coke precursor. The deactivation rate (-da/dt) expression that most satisfactorily correlated with the data was the following

$$-\frac{da}{dt} = \frac{k_d K_{FAL}^2 P_{Furan}^2}{(1 + K_{FAL}P_{Furan} + K_{RS}P_{Furan} + K_{FAL}P_{FAL})^2} a^{1.5} \tag{7}$$

This expression was derived from the mechanism involving two adjacent adsorbed furfural molecules. It was found that resulting values for the equilibrium constants of the adsorption of furfural that yields the coke precursor (K_{FAL*}) are significantly higher than the equilibrium adsorption constant for furfural (K_{FAL}) used in the reaction kinetics. This enhancement would indicate that deactivation is enhanced by adsorbate-adsorbate interactions that strengthen the adsorption of furfural. Interestingly, the activation energy for deactivation derived from the fitting was 6.7 kcal/mol, which is much lower than that of the main reaction (14.8 kcal/mol). The former is associated with a condensation step, the latter with a C–C bond cleavage, which requires significant energy.

7.3.5 OPENING OF THE FURANYL RING

Ring opening is another potentially important reaction in the conversion of furfural over metal supported catalysts. Several pathways have been proposed for the ring opening, as shown in Figure 13. It can be seen that the products obtained from this reaction are predominantly alcohols, which could be promising biofuel components, such as butanol. It may also be used as valuable intermediates in the production of chemicals. For example, 1,5- pentanediol can be used as a monomer in the production of polyesters and polyurethanes [133]. At high enough temperatures and under acidic conditions, 1,5-pentanediol can be dehydrated to pentanes, which has a reasonable octane number (ON), but an unacceptably high vapor pressure to be incorporated in gasoline [134].

Several metal catalysts could be used to catalyze the ring opening reaction. Some of the metal catalysts mentioned in the literature include Ni [112,134,135], Rh [136], Pt[137] and Cu-chromite [103]. It is interesting to point out that by changing the catalyst one can obtain different alcohols since different catalysts have different selectivities towards the various possible reaction paths, particularly the first step. For example, 1,5-pentanediol and 2- pentanol were observed when hydrogenation and hydrogenolysis, respectively, occurred as the first step in the reaction sequence, while butanol was produced when the reaction started with the decabonylation of furfural. A 33 % yield of 2-pentanol and 2-pentanone was achieved at 300 °C over a catalyst containing Cu/Cr/Ni/Zn/Fe with molar ratios of 43:45:8:3:1 [103]. In this case, the hydrogenolysis of furfural to 2-methyl furan was reported as the primary reaction. In contrast, butanal, butanol, and butane, which are derived from the ring opening of furan were observed as the main products from the conversion of furfural over Ni/SiO_2 catalyst at > 200 °C since furfural decarbonylation to furan is readily catalyzed by Ni at these high temperatures, as mentioned above [112]. On the other hand, at lower temperatures (140 °C) hydrogenation of furfural to furfuryl alcohol is preferred over Ni catalyst. As a result, the ring opening produces mainly 1,5- pentandiol, which as mentioned can be later dehydrated to C_5-hydrocabons over an acidic support (SiO_2-Al_2O_3) Ref. [134].

Finally, Rh has also been used to catalyze the ring opening reactions. For example, the ring opening of tetrahydrofurfuryl alcohol was observed on Rh/SiO_2 after hydrogenation of furfuryl alcohol in aqueous solution [136]. Without any modification, Rh/SiO_2 showed very low activity for this reaction and the main product was 1,2-pentanediol (66 % selectivity). However, after addition of MoOx, a remarkable increase in activity was observed and the product selectivity was observed to change from 1,2-pentanediol to 1,5- pentanediol (> 93 % selectivity), as shown in Figure 14. This interesting change in behavior might have been explained in terms of an enhancement of adsorbate interaction with the catalysts via the OH group of the molecule and the MoOx species. According to this proposal, tetrahydrofurfuryl alcohol is adsorbed on MoOx species via the OH group and the cleavage of the C–O bond is then catalyzed by the neighboring Rh particles.

7.4 MOLECULAR ENGINEERING OF FUEL COMPONENTS DERIVED FROM FURFURALS

A quick examination of the different furfural conversion strategies reviewed above indicates that by choosing the right catalyst and reaction conditions the researcher can pick and choose from a broad range of potential fuel components. As summarized in Figure 15, starting with furfural, one can start with hydrogenation on a metal (Cu, Pd, Ni, Pt). The product furfuryl alcohol would not be the most desirable fuel component since it is fully miscible in water (see Table 2). If the furanyl ring is further hydrogenated and then the ring opened (e.g. on Pd and Ni, respectively), 1,5 pentanediol can be obtained. As shown in Table 2, this diol has a very good octane number, but its high solubility in water may also hinder its applications as a fuel component, similar to furfuryl alcohol. A similar situation may occur with butanol, which can be relatively easily obtained with a combination of decarbonylation/hydrogenation and ring opening (on Ni). While this potential fuel component has a relatively good octane number and vapor pressure, its water solubility is also high. However, it must be noted that, in this sense, butanol appears much better than ethanol, a fuel component widely used today despite its high water solubility. Two interesting alcohols that can be obtained from furfural and methyl furfural via hydrogenolysis followed by ring opening are pentanol and hexanol, respectively. While their octane numbers are not as high as those of the shorter alcohols, their vapor pressures and water solubilities are very attractive for gasoline components. This is an example of the fuel component analysis that can be done with the molecular engineering approach, in which the researcher may have the possibility of choosing reaction path that optimize several fuel properties simultaneously.

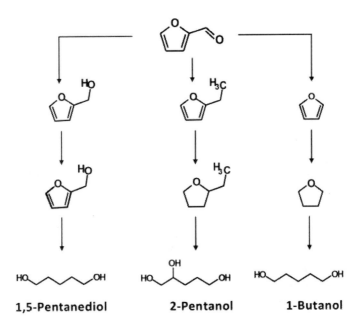

FIGURE 13: Reaction scheme for the ring opening of furfural on metals.

Furfurals as Chemical Platform for Biofuels Production

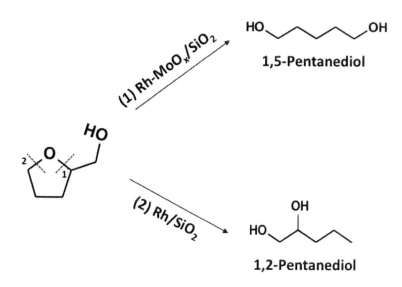

FIGURE 14: Ring opening reaction of tetrahydrofurfuryl alcohol over Rh/SiO$_2$ and Rh-MoOx/SiO$_2$ [136].

Avoiding ring opening produces furan and methyl furan, aromatic compounds that have remarkably high octane numbers and low solubility in water. The only serious limitation of these compounds that may hinder their use as fuel components is their high vapor pressure, which is much higher than those of their corresponding ring-opening derivatives.

Finally, the path that combines aldol-condensation with hydrogenation and hydrodeoxygenation appears greatly attractive for the production of diesel fuel components. As shown in Figure 15, while the C$_8$ compound that results from the condensation of one molecule of acetone with one molecule of furfural is in the gasoline range, the C$_{13}$ compound that results from two furfurals and one acetone is in the diesel range and has a high cetane number, even after hydrodeoxygenation, resulting in an appealing alternative for green diesel fuel.

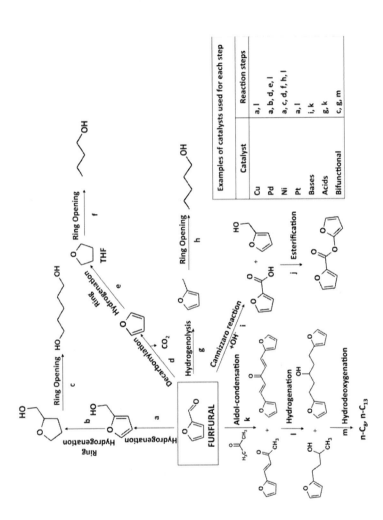

FIGURE 15: Catalytic strategies for the conversion of furfural to fuel components and chemicals. The possible reaction steps are indicated with lowercase letters and the corresponding catalysts listed in the inset.

TABLE 2: Summary of the possible reactions taking place in the conversion of furfural (FAL) and hydroxymethylfurfural (HMF), with the respective products and their properties.

Feedstock	Reactions	Products	ON or CN*	RT Vapor Pressure (mm Hg)	Water solubility (g/l)
FAL	Hydrogenation	furfuryl alcohol	134	3	Fully miscible
FAL	Hydrogenation + Ring opening	1,5-pentane-diol	69	0.02	Fully miscible
FAL	Decarbonylation	furan	109	490	0.0
FAL	Hydrogenolysis	methyl furan	131	140	7
FAL	Decarbonylation + Ring opening	butanol	96	6	77
FAL	Hydrogenolysis	pentanol	57	1.8	22
HMF	Ring opening	hexanol	55	0.93	8.2
HMF	Hydrogenolysis	dimethylfuran	119	26	2.3
HMF	Aldol-condensation	4-2-(furyl)buten-3-en-2-one	128	–	–
HMF	Aldol-condensation	1,5-di-2-furyl pentan-1,4-dien-3-one	75*	–	–
HMF	Aldol-condensation + Hydrogenation	1,5-di-furyl-pentan-3-ol	100*	–	–

In this contribution we have reviewed the possible reaction paths that could be followed for upgrading furfurals (e.g. 2-furfuraldehyde and 5-hydroxymethyl furfural), intermediates from biomass to fuel components and chemicals. It is concluded that by varying the catalyst composition (metal, additives, support), as well as reaction conditions (pressure, temperature, liquid/vapor phase, etc.), one can obtain a rich variety of products with different fuel properties.

REFERENCES

1. Schmidt, L.D.; Dauenhauer, P.J. Nature 2007, 447, 914–915.
2. Rostrup-Nielsen, J.R. Science 2005, 308, 1421–1422.

3. Somerville, C.; Youngs, H.; Taylor, C.; Davis, S.C.; Long, S.P. Science 2010, 309, 790-792.
4. Wyman, C.E.; Hinman, N.D. Appl. Biochem. Biotechnol. 1990, 24/25, 735–753.
5. Wyman, C.E. Appl. Biochem. Biotechnol. 1994, 45/46, 897–915.
6. Tyson, K.S. Fuel Cycle Evaluations of Biomass-Ethanol and Reformulated Gasoline; Report No. NREL/TP-263-2950, DE94000227, National Renewable Energy Laboratory: Golden, CO, 1993.
7. Lynd, L.R.; Cushman, J.H.; Nichols, R.J.; Wyman, C.E. Science 1991, 251, 1318–1323.
8. U.S. Department of Energy, Feedstock Composition Gallery, Washington, DC, 2005; http://www.eere.energy.gov/biomass/feedstock_glossary.html.
9. Perlack, R.D.; Wright, L.L.; Turhollow, A.; Graham, R.L.; Stokes, B.; Erbach, D.C. Biomass as Feedstock for a Bioenergy and Bioproducts Industry: The Technical Feasibility of a Billion-Ton Annual Supply, Report No. DOE/GO-102995-2135; Oak Ridge National Laboratory: Oak Ridge, TN, 2005; http://www.osti.gov/ bridge.
10. Klass, D.L. Biomass for Renewable Energy, Fuels and Chemicals; Academic Press: San Diego, 1998.
11. Energy Information Administration Annual Energy Outlook 2005; Report. No. DOE/EIA-0383; U.S. Department of Energy: Washington, DC, 2006; http://www.eia.doe.gov.
12. Aye, M.M.S.; Zhang M. Chem. Eng. Sci. 2005, 60, 6702–6717.
13. Katzer, J.R. Interface Challenges and Opportunities in Energy and Transportation. Energy and Transportation: Challenges for the Chemical Sciences in the 21st Century, The National Academy Press 2003.
14. Lange, J.P. Biofuels Bioprod. Bioref. 2007, 1, 39–48.
15. Pasadakis, N.V.; Gaganis, V.; Foteinopoulos, C. Fuel Proc. Technol. 2006, 87, 505–509.
16. Resasco, D.E.; Crossley, S., AIChE J. 2009, 55, 1082–1089.
17. Santana, R.C.; Do, P.T.; Alvarez, W.E.; Taylor, J.D.; Sughrue, E.L.; Resasco, D.E. Fuel 2006, 85, 643–656.
18. Taylor, J.; McCormick, R.; Clark, W. Relationship between Molecular Structure and Compression Ignition Fuels, both Conventional and HCCI. August 2004 NREL Report on the MP-540-36726, Non-Petroleum-Based Fuels.
19. Do, P.; Crossley, S.; Santikunaporn ,M.; Resasco, D.E. Catalytic Strategies for Improving Specific Fuel Properties. Catalysis: Specialist Periodical Reports. The Royal Society of Chemistry, London 2007, 20, 33–61.
20. Crossley, S.; Alvarez, W.E.; Resasco, D.E. Energy Fuels. 2008, 22, 2455–2464.
21. Do, P.T.; Alvarez, W.E.; Resasco, D.E. J. Catal. 2006, 238, 477–488.
22. Lichtenthaler, F.W. Acc. Chem. Res. 2002, 35, 728–737.
23. Lewkowski, J. ARKIVOC 2001, 17–54.
24. Green, J.H.S.. Revision of the Values of the Heats of Formation of Normal Alcohols, Chem. Ind. (London), 1960, 1215–1216.
25. Xing, R.; Subrahmanyam, A.V.; Olcay, H.; Qi, W.; Van Walsum, J.P.; Pendse, H. Huber, G.W. Green Chem. 2010, 12, 1933–1946.
26. Hall, D.O.; Rao, K.K. Photosynthesis (Studies in Biology). Cambridge University Press 1999, 6th Ed.

27. Sinnott, M.L. Carbohydrate Chemistry and Biochemistry Structure and Mechanism, The Royal Society of Chemistry 2007.
28. Wyman, C.E.; Decker, S.R.; Himmel, M.E.; Brady, J.W.; Skopec, C.E.; Viikari, L. In Polysaccharides; Dumitriu, S., Ed.; Marcel Dekker: New York, 2005, 2nd Ed.
29. Public Law 110–140. Energy Independence and Security Act, Dec. 2007.
30. Ek, M.; Gellerstedt, G.; Henriksson, G. Wood Chemistry and Biotechnology. Ed. De Gruyter, 2009, 121–145.
31. Lewis, N.G.; Sarkanen, S. Lignin and Lignan Biosynthesis, Vol. 697, 1998 American Chemical Society.
32. Sun, Y.; Cheng, J. Bioresour. Technol. 2002, 83, 1–11.
33. Mosier, N.; Wyman, C.; Dale, B.; Elander, R.; Lee, Y.Y.; Holtzapple, M.; Ladisch, M. Bioresour. Technol. 2005, 96, 673–686.
34. Azzam, A.M. J. Environ. Sci. Health. B. 1989, 24, 421–433.
35. Huber, G.W.; Dumesic, J.A. Catal. Today 2006, 111, 119–132.
36. Bridgwater, A.V. Fuel 1995, 14, 631–653.
37. Sutton, D.; Kelleher, B.; Ross, J.R.H. Fuel Process. Technol. 2001, 73, 155–173.
38. Mohan, D.; Pittman, C.U.; Steele, P.H. Energy Fuels 2006, 20, 848–889.
39. Demirbas, A. Prog. Energy Combust. Sci. 2007, 33, 1–18.
40. Van Steen, E.; Claeys, M. Chem. Eng. Technol. 2008, 31, 655–666.
41. Bridgwater, A.V. Therm. Sci. 2004, 8, 21–49.
42. Elliott, D.C.; Schiefelbein, G.F. Amer. Che,. Soc. Div. Fuel Chem. Preprints 1989, 34, 1160–1166.
43. Vispute, T.P.; Zhang, H.; Sanna, A.; Xiao, R.; Huber, G.W. Science 2010, 330, 1222–1227.
44. Gangadharam, A.; Shen, M.; Sooknoi, T.; Resasco, D.E.; Mallinson, R.G. Appl. Catal. A 2010, 385, 80–91.
45. Zhu, X.; Lobban, L.L.; Mallinson, R.G.; Resasco, D.E. J. Catal. 2010, 271, 88–98.
46. Pham, T.; Lobban, L.L.; Resasco, D.E.; Mallinson, R.G.; J. Catal. 2009, 266, 9–14.
47. Peralta, M.A.; Sooknoi, T.; Danuthai, T.; Resasco, D.E. J. Molec. Catal. A 2009, 312, 78–86.
48. Ausavasukhi, A.; Sooknoi, T.; Resasco, D.E. J. Catal. 2009, 268, 68–78.
49. Ladish, M.R.; Lin, K.W.; Voloch, M.; Tsao, G.T. Enzyme Microb. Technol. 1983, 5, 82–102.
50. Diebold, J.P. A Review of the Chemical and Physical Mechanisms of the Storage Stability of Fast Pyrolysis Bio-Oils. 2000. Thermalchemie, Inc. Lakewood, Colorado National Renewable Energy Laboratory http://www.doe.gov/bridge
51. Adjaye, J.D.; Katikaneni, S.P.R.; Bakhshi, N.N. Fuel Process. Technol. 1996, 48, 115–143.
52. Sharma, R.K.; Bakhshi, N.N. Energy Fuels 1993, 7, 306–314.
53. Katikaneni, S.P.R.; Adjaye, J.D.; Bakhshi, N.N. Energy Fuels 1995, 9,1065–1078.
54. Roman-Leshkov, Y.; Barrett, C.J.; Liu, Z.Y.; Dumesic, J.A. Nature 2007, 447, 982–985.
55. Crossley, S.; Faria, J.; Shen, M.; Resasco, D.E. Science 2010, 327, 68–72.
56. Fukuda, H.; Kondo, A.; Noda, H. J. Biosci. Bioeng. 2001, 92, 405–416.
57. Kirubakaran, V.; Sivaramakrishnan, V.; Nalini, R.; Sekar, T.; Premalatha, M.; Subramanian, P. Renew. Sustain. Energ. Rev. 2009, 13, 179–186.

58. Alauddin, Z.A.B.Z.; Lahijani, P.; Mohammadi, M.; Mohamed, A.R. Renew. Sustain. Energ. Rev. 2010, 14, 2852–2862.
59. The Conversion of Wood and other Biomass to Bio-oil; ENSYN Group, Inc.: Greely, CA, June 2001.
60. Czernik, S.; Bridgwater, A.V. Energy Fuels 2004, 18, 590–598.
61. Peacocke, G.V.C.; Russel, P.A.; Jenkins, J.D.; Bridgwater, A.V. Biomass Bioenergy 1994, 7, 169–178.
62. Piskorz, J.; Scott, D.S.; Radlien, D. Composition of Oils Obtained by Fast Pyrolysis of Different Woods. In Pyrolysis Oils from Biomass: Producing Analyzing and Upgrading; American Chemical Society: Washington, DC, 1988; pp 167–178.
63. Bridgwater, A.V.; Peacocke, G.V.C. Energy Rev. 2000, 204, 117–126.
64. Antal, M.J.; Varhegyi, G. Ind. Eng. Chem. Res. 1995, 34, 703–717.
65. Hallen, R.T.; Sealock, L.T.; Cuello, R.; Bridgwater, A.V. In Research in Thermochemical Biomass Conversion; Kuester, J. L., Ed.; Elsevier: London, UK, 1988.
66. Patwardhan, P.R.; Satrio, J.A.; Brown, R.C.; Shanks, B.H. Biores. Technol. 2010, 101, 4646–4655.
67. Patwardhan, P.R.; Satrio, J.A.; Brown, R.C.; Shanks, B.H. J. Anal. Appl. Pyrol. 2009, 86, 323–330
68. Ponder, G.R.; Richards, G.N.; Stevenson, T.T. J. Anal. Appl. Pyrol. 1992, 22, 217–229.
69. Yang, C.; Lu, X.; Lin, W.; Yang, X.; Yao, J. Chem. Res. Chinese U. 2006, 22, 524–532.
70. Brownlee, H.J.; Miner, C.S. Ind. Eng. Chem. 1948, 40, 201–204.
71. March, J. Advanced Organic Chemistry, John Wiley & Sons, New York, 1992.
72. Lowry, T.M. J. Chem. Soc. 1927, 2554–2567.
73. Franck, E.U. Z. Phys. Chem. Neue Folge 1956, 8, 107–126.
74. Zeitsch, K.J. The Chemistry and technology of furfural and its many by-products. Elsevier Science B.V. Amsterdam, 2000.
75. Moliner, M.; Roman-Leshkov, Y.; Davis, M.E. PNAS 2010, 107, 6164–6168.
76. Hurd, C.D.; Isenhour, L.L. J. Amer. Chem. Soc. 1932, 54, 317–330.
77. Zhao, H.; Holladay, J.E.; Brown, H.; Zhang, Z.C. Science 2007, 316, 1597–1600.
78. Lima, S.; Neves, P.; Antunes, M.M.; Pillinger, M.; Ignatyev, N.; Valente, A.A. Appl. Catal. A 2009, 363, 93–99.
79. Binder, J.B.; Raines, R.T. J. Am. Chem. Soc. 2009, 131, 1979–1985.
80. Lai, L.; Zhang, Y. ChemSusChem 2010, 3, 1257–1259.
81. Chheda, J.N.; Dumesic, J.A. Catal. Today 2007, 123, 59–70.
82. Barrett, C.J.; Chheda, J.N.; Huber, G.W.; Dumesic, J.A. Appl. Catal. B 2006, 66, 111–118.
83. Casanova, O.; Iborra, S.; Corma, A. J. Catal. 2010, 275, 236–242.
84. Kunkes, E.L.; Simonetti, D.A.; West, R.M.; Serrano-Ruiz, J.C.; Gartner, C.A.; Dumesic, J.A. Science. 2008, 322, 417–421.
85. Simonetti, D.A.; Dumesic, J.A. ChemSusChem 2008, 1, 725–733.8
86. West, R.M.; Liu, Z.Y.; Peter, M.; Dumesic, J.A. ChemSusChem 2008, 1, 417–424.
87. Chheda, J.N.; Roman-Leshkov, Y.; Dumesic, J.A. Green Chem. 2007, 9, 342–350.
88. Chheda, J.N.; Huber, G.W.; Dumesic, J.A. Angew. Chem. Int. Ed. 2007, 46, 7164–7183.

89. Kelly, G.J.; Jackson, S.D. Aldol Condensation of Aldehydes and Ketones over Solid Base Catalysts. Catalysis in Application, Jackson, S.D.; Hargreaves, J.S.J.; Lennon, D. Royal Society of Chemistry (Great Britain).
90. Serrano-Ruiz, J.C.; Braden, D.J.; West, R.M.; Dumesic, J.A. App. Catal. B 2010, 100, 184–189.
91. Kunkes, E.L.; Gürbüz, E.I.; Dumesic, J.A. J. Catal. 2009, 266, 236–249.
92. Huber, G.W.; Chheda, J.N.; Barrett, C.J.; Dumesic, J.A. Science 2005, 308, 1446–1449.
93. West, R.M.; Kunkes, E.L.; Simonetti, D.A.; Dumesic, J.A. Catal. Today 2009, 147, 115–125.
94. Klass, D.L. Organic Commodity Chemicals from Biomass, Ref. 2, 495–546.
95. Carbohydrates as Organic Raw Materials I, Ed. Lichtenthaler, F.W. VCH, Weinheim, 1991, 367.
96. Beck, J.S.; Vartuli, J.C.; Roth, W.J.; Leonowicz, M.E.; Kresge, C.T.; Schmitt, K.D.; Chu, C.T.W.; Olson, D.H.; Sheppard, E.W.; McCullen, S.B.; Higgins, J.B.; Schlenkert, J.L. J. Am. Chem. Soc. 1992, 114, 10834–10843.
97. Merck Company, Inc., GB 887360, 1962.
98. Cram, D.J. DE 2539324, 1976.
99. Chundury, D.; Szmant, H.H. Ind. Eng. Chem. Prod. Res. Dev. 1981, 20, 158–163.
100. Wen, R.; Yu, F.; Dong, X.; Miao, Y.; Zhou, P.; Lin, Z.; Zheng, J.; Wang, H.; Huang, L.; Qing, D. CN 1456556, 2003.
101. Seo, G.; Chon, H. J. Catal. 1981, 67, 424–429.
102. Rao, R.; Dandekar, A.; Baker, R.T.K.; Vannice, M.A. J. Catal. 1997, 171, 406–419.
103. Zheng, H.Y.; Zhu, Y.L.; Teng, B.T.; Bai, Z.Q.; Zhang, C.H.; Xiang, H.W.; Li, Y.W. J. Mol. Catal. A 2006, 246, 18–23.
104. Nagaraja, B.M.; Padmasri, A.H.; David Raju, B.; Rama Rao, K.S. J. Mol. Catal. A 2007, 265, 90–97.
105. Reddy, B.M.; Reddy, G.K.; Rao, K.N.; Khan, A.; Ganesh, I. J. Mol. Catal. A 2007, 265, 276–282.
106. Huang, W.; Li, H.; Zhu, B.; Feng, Y.; Wang, S.; Zhang, S. Ultrason. Sonochem. 2007, 14, 67–74.
107. Sitthisa, S.; Sooknoi, T.; Ma, Y.; Balbuena, P.B.; Resasco, D.E. J. Catal. 2011, 277, 1–13.
108. Claus, P. Top. Catal. 1998, 5, 51–62.
109. Li, H.; Luo, H.; Zhuang, L.; Dai, W.; Qiao, M. J. Mol. Catal. A. 2003, 203, 267–275.
110. Liaw, B.; Chiang, S.; Chen, S.; Chen, Y. Appl. Catal. A 2008, 346, 179–188.
111. Kijeński, J.; Winiarek, P.; Paryjczak, T.; Lewicki, A.; Mikolajska, A. Appl. Catal. A 2002, 233, 171–182.
112. Sitthisa, S.; Resasco, D.E. Catal. Lett. 2011, 141, 784–791.
113. Sitthisa, S.; Pham, T.; Prasomsri, T.; Sooknoi, T.; Mallinson, R.G.; Resasco, D.E. J. Catal. 2011, 280, 17–27.
114. Bradley, M.K.; Robinson, J.; Woodruff, D.P. Surf. Sci. 2010, 604, 920–925.
115. Avery, N.R. Surf. Sci. 1983, 125, 771–786.
116. Shekhar, R.; Plank, R.V.; Vohs, J.M.; Barteau, M.A. J. Phys. Chem. B 1997, 101, 7939–7951.
117. Madix, R.J.; Yamada, T.; Johnson, S.W. Appl. Surf. Sci. 1984, 19, 43–58.

118. Vannice, M.A. In "Kinetics of Catalytic Reactions" Springer (2005).
119. Rioux, R.M.; Vannice, M.A. J. Catal. 2003, 216, 362–369.
120. Rojas, H.; Martinez, J.J.; Reyes, P. Dyna 2010, 163, 151–159.
121. Vaidya, P.D.; Mahajani, V.V. Ind. Eng. Chem. Res. 2003, 42, 3881–3885.
122. Rojas, H.; Borda, G.; Rosas, D.; Martinez, J.J.; Reyes, P. Dyna 2008, 155,115–122.
123. Singh, H.; Prasad, M.; Srivastava, R.D. J. Chem. Technol. Biotechnol. 1980, 30, 293–296.
124. Srivastava, R.D.; Guha, A.K. J. Catal. 1985, 91, 254–262.
125. Jung, K. J.; Gaset, A.; Molinier, J. Biomass 1988, 16, 63–76.
126. Moshkin, P.A.; Preobrazhenskaya, E.A.; Berezina, B.B.; Markovich, V.E.; Kudrina, E.G.; Papsueve, V.P.; Sokolina, E.A. Khimiya Oeterotsiklicheskikh Soedinenii 1966, 2, 3–7.
127. Coca, J.; Morrondo, E.S.; Sastre, H. J. Chem. Technol. Biotechnol. 1982, 32, 904–908
128. Coca, J.; Morrondo, E.S.; Parra, J.B.; Sastre, H. React. Kinet. Catal. Lett. 1982, 20, 3–4.
129. Wambach, L.; Irgang, M.; Fischer, M. (BASF Aktiengesellschaft), US patent 4,780,552 (1988).
130. Zhang, W.; Zhu, Y.; Niu, S.; Li, Y. J. Mol. Catal A: Chem (2011) In press.
131. Lopez, N.; Norskov, J.K. Surf. Sci. 2001, 477, 59–75.
132. Froment, G.F.; Hosten, L.H. Catalysis Science and Technology (Anderson, J.R.; Boudart, M. Eds.) Springer-Verlag, New York, 1981.
133. Schlaf, M. Dalton Trans. 2006, 4645–4653.
134. Xinghua, Z.; Tiejun, W.; Longlong, M.; Chuangzhi, W. Fuel 2010, 89, 2697–2702.
135. Vetere, V.; Merlo, A.B.; Ruggera, J.F.; Casella, M.L.; Braz, J. Chem. Soc. 2010, 21, 914–920.
136. Koso, S.; Ueda, N.; Shinmi, Y.; Okumura, K.; Kizuka, T.; Tomishige, K. J. Catal. 2009, 267, 89–92.
137. Kliewer, C.J.; Aliaga, C.; Bieri, M.; Huang, W.; Tsung, C.; Wood, J.B.; Komvopoulos, K.; Somorjai, G.A. J. Am Chem. Soc. 2010, 132, 13088–13095.

PART V

INCINERATION AND CARBONIZATION

CHAPTER 8

Incineration of Pre-Treated Municipal Solid Waste (MSW) for Energy Co-Generation in a Non-Densely Populated Area

ETTORE TRULLI, VINCENZO TORRETTA, MASSIMO RABONI, AND SALVATORE MASI

8.1 INTRODUCTION

The main objective in integrated solid waste management (ISWM) [1,2] is to implement technologies that reduce the environmental pressure by recovering both the fractions with a considerable value on the market and the non-traditional ones (e.g., organic [3,4,5,6,7,8], medical [9,10,11], automotive shredder residues [12], WEEE [13]). Moreover, any good management system also includes the involvement of the people, who have to be aware of the environmental benefits and of the reduced danger to health that results from a correct behavior [14,15]. Such an objective is highlighted by European Union (EU) legislation, which produced several Directives on waste disposal, treatment and incineration [16,17,18,19].

Incineration of Pre-Treated Municipal Solid Waste (MSW) for Energy Co-Generation in a Non-Densely Populated Area. © Trulli E, Torretta V, Raboni M, and Masi S. Sustainability 5,12 (2013), doi:10.3390/su5125333. Licensed under a Creative Commons Attribution 3.0 Unported License, http://creativecommons.org/licenses/by/3.0/.

Such Directives: (i) prohibit waste recovery and disposal that have a negative impact on both the human health and the environment; (ii) aim at the reduction of waste production as well as the promotion of the reuse, the recycling and the recovery activities. The Italian Government acknowledged these Directives [20,21], also imposing the energy recovery from waste incineration.

According to the above-mentioned regulations and principles, the ISWM should both reduce landfilling and increase energy and materials recovery in order to lower environmental impact, energy resources consumption and economic costs. For example, landfilling of energy-rich waste should be avoided as far as possible, partly because of the negative environmental impacts of the technique, but mainly because of the low resources recovery [22].

Various types of Life-Cycle-Analysis (LCA) have been proposed for determining the most environmental-sound ISWM procedure [23,24]. Most of them focus on high percentages of separated collection in relatively small and densely populated area: some examples are described in [25,26,27,28]. However, in a scarcely-populated area, the environmental (e.g., GHG emissions) and economic impact of MSW collection is high because of fuel consumption [25,29,30]. The ISWM issue in a non-densely populated area concerns developing and developed countries. For example, 60% of EU surface has a population density of less than 100 inhab km^{-2} [31].

In order to reduce the landfill volumes as well as to close the MSW cycle, a solution can be the waste incineration with energy recovery [32]. Therefore, excluding direct landfilling of MSW, the solutions for ISWM are essentially three (Figure 1): direct burning of the raw MSW; accelerated stabilization of the whole MSW prior to incineration; MSW mechanical pre-treatment with secondary fuel (or RDF) production prior to incineration and organic matter aerobic stabilization (the so-called biostabilization) before landfilling.

Regarding the last solution, the mechanical pre-treatment of waste by sieving has aroused great interest, because it influences the MSW volumes addressed for both incineration and landfilling [33]; as consequence, the process reduces the environmental impact of the whole ISWM system. Sieving is carried out on raw waste and it allows separating out a flow of material that is characterized by a higher energy content (heating value, HV, and net calorific value, NCV) [34]. The separation leads to a reduction in the combustion section, even if it partly penalizes the energy recovery. Moreover, in plants which treat waste coming from basins with different waste management policies, the pre-treatment process allows to guarantee a better quality of secondary fuel.

The case study presented in this paper regards the methodological approach applied to determine which technical solution may be proposed for solving some ISWM issues in a non-densely populated area, with specific reference to Basilicata, an Italian region. The analysis of the MSW composition and quantity (current situation and future trend) has been carried out. The results, combined with the waste size characterization, allowed to define the optimal mechanical pre-treatment in order to achieve an optimal balance between GHGs emissions and heat-and-power production.

8.2 MATERIALS AND METHODS

8.2.1 INVESTIGATED AREA

Basilicata is a predominantly mountainous region (Figure 2a) that covers about 10,000 km^2 in the center of Southern Italy. The population is less than 580,000 inhabitants [35]. Most of the 131 districts have a population below 3000 inhab with an average population density varying between 31 and 380 inhab km^{-2} (average: 57 inhab km^{-2}; Figure 2b).

The average MSW production (1.08 kg inhab^{-1} d^{-1}) is less than the national average [36,37]. During the last few years, a 2% y^{-1} increase in the MSW average production occurred, with peaks of 5% y^{-1} in the largest towns.

8.2.2 CHARACTERISTICS OF WASTE PRODUCTION

A survey was carried out with the aim of determining both the sieved waste composition and the influence on the combustible waste characteristics [37]. The mesh size was determined by considering that it has a considerable effect on the quantity of the obtainable energy [38].

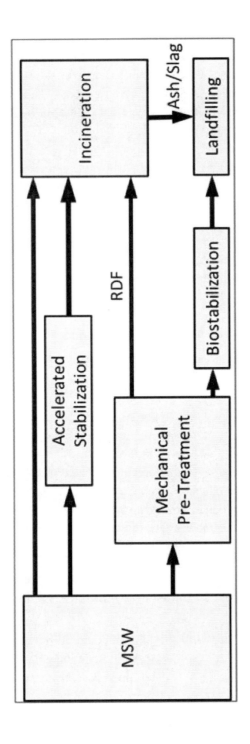

FIGURE 1: Typical municipal solid waste (MSW) treatment options.

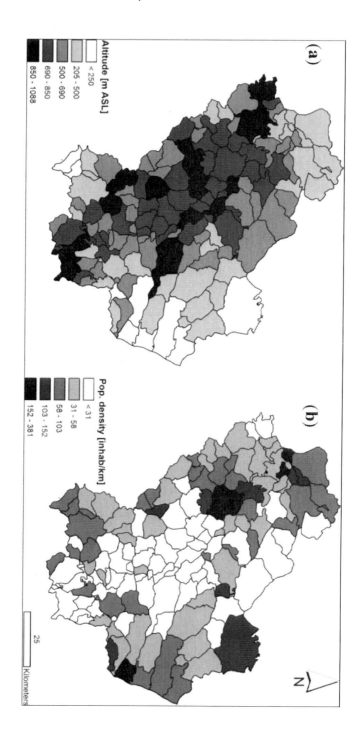

FIGURE 2: Average altitude (a) and population density (b) in Basilicata region.

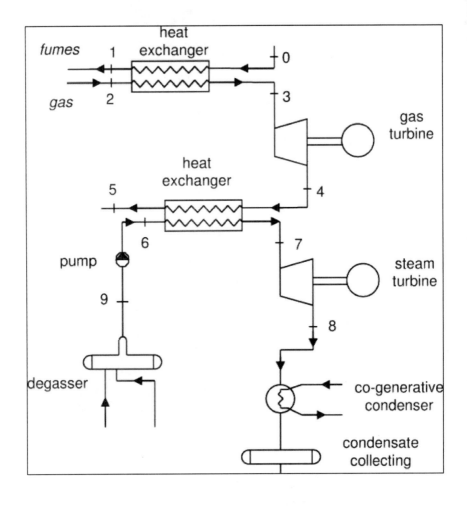

FIGURE 3: The examined combined cycle plant layout. Numbers indicates the calculation sections.

Experimental tests were carried out on the waste collected in urban centers, with a production rate that fell in the range 0.8–1.2 kg inhab^{-1} d^{-1}. The percentages of separation by sieving were determined experimentally and the different product fraction percentages were deduced on the basis of the average composition of the waste. The data was determined as a function of the amount, in terms of weight, of the individual fractions of over-sieved (OS) and under-sieved (US) waste. This data can be extended to all compositions, on the assumption that if the amount, in terms of weight, of the single fraction is varied, the size does not change. Predictions in MSW evolution concerning quantity and composition were carried out considering: (i) past data; (ii) population growth; (iii) evolution of policies regarding separated collection [36].

The HV of each fraction was estimated using data found in scientific literature [32,39] (organic: 2,930.2 kJ kg^{-1}; paper: 12,558.0 kJ kg^{-1}; plastic and rubber: 20,930.0 kJ kg^{-1}; wood, textile and leather: 15,488.2 kJ kg^{-1}; under-sieved waste below 20 mm: 5,651.1 kJ kg^{-1}) and was applied to each OS fraction in order to obtain the respective energy content.

TABLE 1: Emission factors adopted for the environmental effects assessment.

Treatment		GHGs emissions	
		Unit	Value
Biostabilization	without energy recovery	kg CH$_4$eq kg^{-1} biodegradable VSS	2.5
	with energy recovery		1.5
Incineration plant	50% of biodegradable fraction removal	kg CH$_4$eq kg^{-1} VSS	1.5
Landfilling (50% of biogas capture)	without energy recovery	kg CH$_4$eq kg^{-1} biodegradable VSS	10.5
	with energy recovery		9.5

8.2.3 ENVIRONMENTAL EFFECTS ASSESSMENT

The GHGs emission assessment of three ISWM possible solution was carried out considering the influence of the sieve cut-off [40]. Such solutions are: (a) the direct MSW landfilling; (b) US landfilling and OS incineration; (c) US biostabilization and OS incineration. Table 1 reports the emission factors assumed for the environmental assessment [40,41].

8.2.4 MODELING

8.2.4.1 PROCESS AND MODEL DESCRIPTION

The combined gas-steam power cycle plant (Rankine-Brayton cycle [42]) for power and heat production may be regarded as one of the best strategies for MSW management, in terms of efficiency, pollution and management costs [43,44,45]. The layout of the examined plant is shown in Figure 3.

The thermal cycle used for power recovery is composed of a gas turbine coupled with a water steam cycle, where the heat entering the steam cycle is obtained from the thermal recovery carried out on the gas turbine exhaust. Before being sent for treatment, the high-temperature combustion gases go through a heat exchanger, where heat is transferred, and what comes out is a liquid at a lower temperature. Then the heated and compressed air is sent to the gas turbine. The emitted gases are introduced into a steam generator. A mono-phase counter-pressure steam turbine was considered. The plant is completed by an electrical power generator and by the heat exchangers for producing hot water. Moreover, the plant includes a condensate collector, centrifuge pumps and a degasser for treating the condensate water and the steam. For the gas-gas heat exchanger a Ljungström-type rotating system [46] was used.

In the case of combustion under practical conditions (n > 1), the exhaust gas discharge is equal to:

$$m_f = m_{sw} - (m_{sg} + m_{fa}) + m_{ae} \qquad (1)$$

In order to calculate the above terms, we used the following equations:

$$m_{sg} = \chi_{sg} \cdot m_{sw} \tag{2}$$

$$m_{fa} = \chi_{fa} \cdot m_{sw} \tag{3}$$

$$m_{at} = \chi_{at} \cdot m_{sw} \tag{4}$$

$$m_{ae} = n \cdot m_{at} \tag{5}$$

If there are no unburned substances, and no heat exchange between the gases and the combustion chamber, the gas temperature only depends on the fuel characteristics, the air index n as well as the initial air and fuel temperatures. In practice, a temperature T_f equal to the theoretical combustion temperature, T_{ad}, is considered:

$$T_{ad} = \tau + \frac{HV}{c_f \cdot (n \cdot m_{at})} \tag{6}$$

The temperature of the gas entering the turbine is given by the equation below, by accepting for the gas-gas exchanger an efficiency of ε_{g-g}:

$$T_3 = T_2 + \frac{\varepsilon_{g-g}(m_f - c_f)\Delta T_{0-2}}{m_g \cdot c_g} \tag{7}$$

The temperature of the gas leaving the turbine, considering the 3-4 adiabatic transformation (Figure 3) as reversible, is given by the following equation:

$$T_4 = T_3 \cdot \beta^{(1-k)/k} \tag{8}$$

The electrical power, produced by the alternator coupled to the gas turbine, is given by the equation:

$$W_{GT} = \eta_{GT} \cdot (m_g \cdot c_{GT}) \cdot \Delta T_{3-4} \tag{9}$$

With an efficiency of the air-water heat exchanger of ε_{g-w}, the steam discharge in the turbine is:

$$m_{ST} = \frac{(m_g c_{GT}) \cdot \varepsilon_{g-w} \cdot \Delta T_{4-6}}{\Delta h_{7-6}} \tag{10}$$

Considering a total efficiency ηS_T for the steam turbine section, the electrical power developed is:

$$W_{GT} = \eta_{GT} \cdot m_{ST} \cdot \Delta h_{7-8} \tag{11}$$

The steam/water heat exchanger operates with a gradient of ΔT_{7-6}.
The co-generated thermal energy is equal to:

$$Q = m_s \cdot \Delta h_{8-9} \tag{12}$$

A water discharge is transferred, and is equal to:

$$m_W = \frac{Q}{c_W \cdot \Delta T_{7-6}} \tag{13}$$

The first-principle yield of the co-generative combined cycle plant (η_1) is:

$$\eta_I = \frac{W_{ST} + W_{GT} + Q}{m_{sw} \cdot HV} \tag{14}$$

The yield is lower because a part of Q is wasted while it is transported to the heating.

In co-generative plants, we also consider a second-principle yield (η_{II}) which takes into account that the quantity of electrical power is much greater than the thermal power.

The analyzed energy is intended as the work, which can be obtained as a system returns to steady conditions. In the components of work production, the energy flow coincides with the electrical power; in the condenser, the energy flow is lower than the thermal flow because of the increase in the exchange fluid entropy. The available energy entering the plant coincides with the thermal power produced by the waste combustion. In the case of electricity generation only, η_{II} is equivalent to η_I. In the case of co-generation we have:

$$\eta_{II} = \frac{W_{ST} + W_{GT} + Q - (m_w \cdot T_W \cdot \Delta S)}{m_{sw} \cdot HV} \tag{15}$$

In order to use the waste heating value as a heating source for summer air-conditioning, we consider a value Y for the refrigeration yield; therefore we have:

$$\eta_{I summer} = \frac{W_{ST} + W_{GT} + Y \cdot Q_{sw}}{m_{sw} \cdot HV} \tag{16}$$

8.2.4.2 DATA AND ASSUMPTIONS

The waste-to-energy plant is:

- intended to produce power and low temperature heat for feeding a heating network;
- situated in a strategic area that can be reached from every town through the ordinary communication routes;
- provided with a landfill in order to reduce the transportation of residual waste (e.g., ash, slag).

The HV of the secondary fuel used in the incineration plant results from the MSW analysis. Table 2 shows the values of the parameters used in the calculation.

TABLE 2: Parameters for combustion calculations.

	Parameter		Unit	Value
Coefficients	theoretical combustion air (m_{at}/m_{sw})		-	4.300
	air index (n)		-	2.300
Production rates	slag (χ_{sg})		kg kg^{-1}	0.055
	flying ash (χ_{fa})		kg kg^{-1}	0.188
Specific heats	gases		kJ kg^{-1} K^{-1}	1.260
	gas entering the turbine		kJ kg^{-1} K^{-1}	1.009
	gas leaving the turbine		kJ kg^{-1} K^{-1}	1.165
	steam entering the turbine		kJ kg^{-1} K^{-1}	1.091
Yield	heat exchanger	gas-gas (εg-g)	-	0.950
		gas-water (εg-w)	-	0.700
	Turbine	gas-fed	-	0.730
		steam-fed	-	0.730
	refrigerating machine (Y)		-	0.800

The combustion gas temperature, Tf, is 95 °C.

8.3 RESULTS AND DISCUSSION

8.3.1 WASTE PRODUCTION AND ENERGY POTENTIAL EVOLUTION

Figure 4 shows the estimation of waste production as a function of different percentages of separated waste collection.

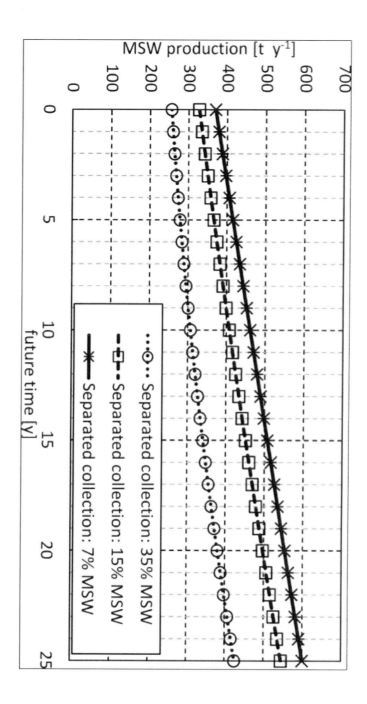

FIGURE 4: Waste production trend as a function of separated waste collection.

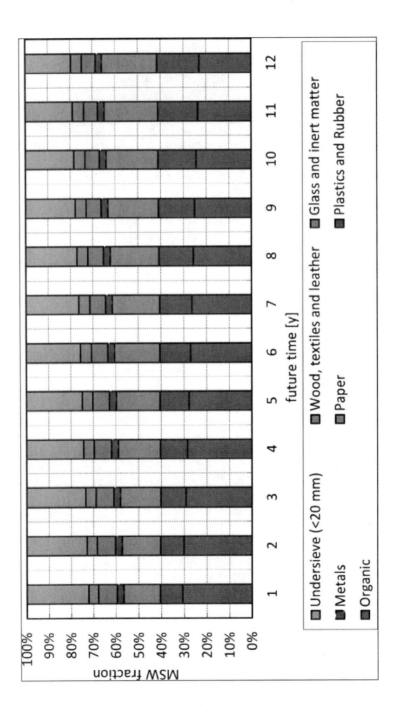

FIGURE 5: Estimation of the MSW composition in Basilicata.

In the next 25 years the MSW production will increase almost linearly, reaching values of 420 and 600 t y^{-1} considering, respectively, the optimistic (35%) and the pessimistic (7%) percentages of separated collection.

Concerning the current composition (Table 3, second column), the MSW production varies greatly. The amount of biodegradable organic waste is equal to about 0.350 kg $inhab^{-1}$ d^{-1}, while the maximum level of paper and plastic collection is 50% of the national average. In the next decade, Basilicata MSW production will be involved in quality variations which will be more remarkable than the quantity ones. Figure 5 shows the results of an estimation of the waste composition in the next 12 y.

A strong reduction of organic fraction (−7.6%), a light reduction of glass (−2.2%), a negligible variation of metals, a sensitive increase of papers and plastics (+8.8% and +8.6%, respectively) and a light increase of wood and textiles (+0.5%) will occur.

Table 3 reports the data relative to the percentage composition of the OS MSW. About half of the waste is smaller than 60 mm and less than 20% is composed by the fine fraction (<20 mm). The composition of the oversized fraction (>120 mm) largely depends on the high HV materials (plastic, paper and textiles), while the major part of the organic fraction is below 60 mm. Glass and inert matter have a homogeneous distribution.

TABLE 3: Percentage composition of both raw and over-sieved (OS) MSW.

MSW fraction	Raw MSW	Sieve cut-off [mm]				
		40	60	80	100	120
Organic	34.3	70.3	39.6	24.3	13.3	6.4
Paper	20.5	96.1	83.5	77.7	69.1	59.4
Plastic and rubber	11.4	93.5	83.2	81.3	73.9	63.4
Wood, textile and leather	5.4	85.3	70.3	63.7	61.9	59.8
Glass and inert matter	6.6	89.7	67.1	50.3	32.4	15.8
Metals	3.0	88.3	82.9	73.3	57.9	49.3
Under-sieved (20 mm)	18.8	-	-	-	-	-
Total	100.0	67.6	50.9	42.5	34.4	27.4

Table 4 shows the HV of separated waste as a function of the sieve size.

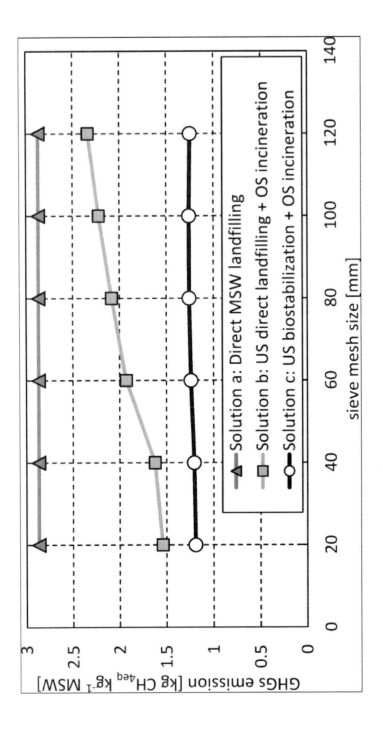

FIGURE 6: Greenhouse gases (GHGs) emissions for different integrated solid waste management (ISWM) solutions.

TABLE 4: Heating value (in kJ kg^{-1}) of the over-sieved MSW fraction.

MSW fraction	Raw MSW	Sieve cut-off [mm]					
		20	40	60	80	100	120
Organic	1,004.0	1,236.6	1,043.4	781.1	574.1	388.4	234.8
Paper	2,576.0	3,172.9	3,659.7	4,225.7	4,709.7	5,177.5	5,591.8
Plastic and rubber	2,378.2	2,929.2	3,287.3	3,887.2	4,549.6	5,112.0	5,510.1
Wood, textile and leather	840.2	1,034.9	1,059.6	1,160.4	1,259.4	1,512.9	1,836.2
Glass and inert matter	-	-	-	-	-	-	-
Metals	-	-	-	-	-	-	-
Under-sieved	5,651.1	1,063.1	-	-	-	-	-
Total	6,861.4	8,373.6	9,050.0	10,054.4	11,092.8	12,190.8	13,172.9

Considering the HV as a function of the MSW sieve cut-off, the trend is almost linear: the energy content of the fine fraction is about 64% of the oversized fraction (>120 mm).

8.3.2 MODELING

8.3.2.1 ENVIRONMENTAL EFFECTS

Figure 6 shows the GHGs emissions for three proposed ISWM solutions.

As expected, the solution with direct landfilling (solution a) generates the highest emissions (e.g., biomethane). In areas where a low waste production occurs (such as Basilicata), the effect is more evident because of the long residence times of waste before the landfill closure and the biogas extraction network completion. The direct US landfilling coupled with OS incineration (solution b) generates more GHGs when the amount of burned MSW diminishes. The sieve cut-off influences also landfill volumes uses and energy recovery. In fact, smaller is the sieve cut-off (20–40 mm), the higher is the waste volume reduction; conversely, higher is the sieve cut-off (60–80 mm), the lower is the waste volume reduction and the higher is the energy recovery. The separation of the wet fraction, which is first aerobically stabilized in reactors and later disposed of in landfills (solution c) is the best solution, independently from the sieve cut-off; moreover, it allows to decrease the incineration unit size (theoretically, partly penalizing the potential energy recovery). We must also take into account that this phase will be gradually reduced by successively boosting of home composting.

8.3.2.2 WASTE INCINERATION AND ENERGY RECOVERY

According to an optimization procedure which considers a compromise between the environmental impact (GHG emissions) and the energy recovery, the sieve cut-off was set equal to 60 mm, corresponding to an HV equal to about 10.1 MJ kg^{-1} (Table 4).

For the gas-gas heat exchanger, the discharge is m_{GT} = 1.3 mf (mf = 60 kg s^{-1}) using a gas temperature in the turbine, T_3 equal to 884 °C. The operating conditions of the gas turbine are: entry pressure of 9.1 kPa; pressure ratio (β) equal to 7.1. The entry temperature and pressure into the turbine are 60 bar and 440 °C, respectively. At the output, there is an optimum level for operation of the turbine, which is no lower than 0.9, and a temperature of approximately 100 °C. With a steam discharge equal to 8.4 kg s^{-1}, the power produced by the alternator coupled with the gas turbine is 25.1 MW. The power developed by the steam turbine is 5.4 MW with a gradient that can be used to produce 30 °C hot water. The co-generated thermal power, Q, is 16.8 MW, which is transferred to a water flow, mw, of 200 kg s^{-1}.

The second-principle yield (η_{II}), in the case of power generation only, is 0.47, while the first- and second-principle yields related to co-generation are 0.74 and 0.54, respectively. In the case of a demand for cold air (used for summer conditioning), $\eta_{I\,summer}$ is 0.68. Therefore, the results show a co-generation second-principle yield lower than η_I because of the difference between the thermal power to the condenser and the electrical power, but higher than ηI in the case of electrical generation only. The major advantage associated with the co-generation respect to the power production must be compared to the higher costs of plant start-up and maintenance.

In the urban area of Potenza (the main town in Basilicata), the locations of the waste disposal, incineration and treatment plants aid the implementation of an interconnection system aimed at energy recovery. Thus, the best solution for the examined case is the integrated management of the energy sources, which entails (i) using the MSW combustible fraction; (ii) managing the landfill and (iii) stabilizing the wastewater treatment plant sludge. In fact, the thermal energy necessary for heating the anaerobic reactors as well as for evaporating percolates and digested sludge can be satisfied by cogeneration plant. Moreover, during summer, when cold is demanded for air-conditioning, it is possible to use an absorption refrigeration cycle, albeit such solution implies an increase in plant costs.

8.4 CONCLUSIONS

The rapid increase in volume and composition of MSW as a result of continuous economic growth, urbanization and industrialization is a problem for national and local governments, if they aim at ensuring an effective and sustainable management of waste. Solid waste incineration requires complex and sophisticated plants, whose running and installation costs are much higher than those of plants that work with traditional fuels. Moreover, energy recovery in fairly small-sized plants is affected by the high cost of the interventions.

The paper focuses on non-densely populated area where the production of urban waste is less than the urban ones. Considering the difficulties in waste collection due to geographic and demographic conditions, separated collection should be carried out only for complying with the regulations. The residual fraction should be sent to the incineration process for energy recovery, with positive environmental effects (in terms of GHGs emissions and landfill volumes) for the whole ISWM system.

Moreover, in low waste production area where the establishment of a unique MSW collection policy is difficult, the residual fraction quality is variable. Such issue can be solved with an appropriate pre-treatment process which improves the characteristics (e.g., heating value) of secondary fuel. The analysis of the obtained results in an Italian non-densely populated area demonstrates the potential of the co-generative incineration residual MSW for energy recovery after sieving, an effective and low cost pre-treatment process. Co-generative incineration is advantageous from an energetic point of view, especially considering the quality and the amount of the energy that can be obtained (power, heat and/or cold).

All these factors, including also the European regulations constraint and the low cost of secondary fuel which tends to increase its heating value, should be taken into account for a sustainable mid-term ISWM system planning in both developing and developed countries where the sustainable management of MSW cycle is penalized by the difficulties in source separated collection and the low flow of material recovery.

REFERENCES

1. United Nations—U.N. Environment Programme. Developing Integrated Solid Waste Management Plan; Training Manual. International Environmental Technology Centre: Osaka/Shiga, Japan, 2009; Volume 4.
2. Cossu, R.; Masi, S. Re-thinking incentives and penalties: Economic aspects of waste management in Italy. Waste Manag. 2013, 33, 2541–2547.
3. Callegari, A.; Torretta, V.; Capodaglio, A.G. Preliminary trial application of biological desulfonation in pig farms' anaerobic digesters. Environ. Eng. Manag. 2013, 12, 815–819.
4. Rada, E.C.; Ragazzi, M.; Torretta, V. Laboratory-scale anaerobic sequencing batch reactor for treatment of stillage from fruit distillation. Water Sci. Technol. 2013, 67, 1068–1074.
5. Martinez, S.L.; Torretta, V.; Minguela, J.V.; Siñeriz, F.; Raboni, M.; Copelli, S.; Rada, E.C.; Ragazzi, M. Treatment of slaughterhouse wastewaters using anaerobic filters. Environ. Technol. 2013.
6. Torretta, V.; Rada, E.C.; Istrate, I.A.; Ragazzi, M. Poultry manure gasification and its energy yield. UPB-Sci. Bull. Ser. D 2013, 75, 231–238.
7. Vaccari, M.; Torretta, V.; Collivignarelli, C. Effect of improving environmental sustainability in developing countries by upgrading solid waste management techniques: A case study. Sustainability 2012, 4, 2852–2861.
8. Hartmann, H.; Ahring, B.K. Anaerobic digestion of the organic fraction of municipal solid waste: Influence of co-digestion with manure. Water Res. 2005, 39, 1543–1552.
9. Chaerul, M.; Tnaka, M. A system dynamics approach for hospital waste management. Waste Manag. 2008, 28, 442–449.

10. Birpmar, M.E.; Bilgili, M.S.; Erdoğan, T. Medical waste management in Turkey: A case study of Istanbul. Waste Manag. 2009, 29, 445–448.
11. Abdulla, F.; Abu Qdais, H.; Rabi, A. Site investigation on medical waste management practices in northern Jordan. Waste Manag. 2008, 28, 450–458.
12. Mancini, G.; Tamma, R.; Viotti, P. Thermal process of fluff: Preliminary tests on a full-scale treatment plant. Waste Manag. 2010, 30, 1670–1682.
13. Torretta, V.; Istrate, I.; Rada, E.C.; Ragazzi, M. Management of waste electrical and electronic equipment in two EU countries: A comparison. Waste Manag. 2013, 33, 117–122.
14. Di Mauro, C.; Bouchon, S.; Torretta, V. Industrial risk in the Lombardy Region (Italy): What people perceive and what are the gaps to improve the risk communication and the partecipatory processes. Chem. Eng. Trans. 2012, 26, 297–302.
15. Morris, M.W.; Su, S.K. Social psychological obstacles in environmental conflict resolution. Am. Behav. Sci. 1999, 42, 1322–1349.
16. European Council. Directive 91/156/EEC Amending Directive 75/442/EEC on Waste; Official Journal L 078, 26/03/1991 P. 0032 – 0037; EEC: Brussels, Belgium, 1991.
17. European Council. Directive 91/689/EEC on Hazardous Waste; Official Journal L 377, 31/12/1991 P. 0020 – 0027; EEC: Brussels, Belgium, 1991.
18. European Parliament and Council. Directive 94/62/EC on Packaging and Packaging Waste; Official Journal L 365, 31/12/1994, P. 0010–0023; EC: Brussels, Belgium, 1994.
19. European Parliament and Council. Directive 2000/76/EC on the Incineration of Waste; Official Journal L 332, 28/12/2000, P. 0091; EC: Brussels, Belgium, 2000.
20. Italian Parliament. Decreto Legislativo 5 febbraio 1997, n. 22 "Attuazione delle direttive 91/156/CEE sui rifiuti, 91/689/CEE sui rifiuti pericolosi e 94/62/CE sugli imballaggi e sui rifiuti di imballaggio". Available online: http://www.parlamento.it/parlam/leggi/deleghe/97022dl.htm (accessed on 2 May 2013).
21. Italian Parliament. Decreto Legislativo 3 aprile 2006, n. 152 "Norme in materia ambientale". Available online: http://www.camera.it/parlam/leggi/deleghe/06152dl.htm (accessed on 2 May 2013).
22. Eriksson, O.; Carlsson, R.M.; Frostell, B.; Björklund, A.; Assefa, G.; Sundquist, J.-O.; Granath, J.; Baky, A.; Thyselius, L. Municipal solid waste management from a systems perspective. J. Clean. Prod. 2005, 13, 241–252.
23. Tascione, V.; Raggi, A. Identification and selection of alternative scenarios in LCA studies of integrated waste management systems: A review of main issues and perspectives. Sustainability 2012, 4, 2430–2442.
24. Bovea, M.D.; Ibáñez-Forés, V.; Gallardo, A.; Colomer-Mendoza, F.J. Environmental assessment of alternative municipal solid waste management strategies: A Spanish case study. Waste Manag. 2010, 30, 2383–2395.
25. De Feo, G.; Malvano, C. The use of LCA in selecting the best MSW management system. Waste Manag. 2009, 29, 1901–1915.
26. Iriarte, A.; Gabarrel, X.; Rieradevall, J. LCA of selective waste collection systems in dense urban areas. Waste Manag. 2009, 29, 903–914.

27. Emery, A.; Davies, A.; Griffiths, A.; Williams, K. Environmental and economic modelling: A case study of municipal solid waste management scenarios in Wales. Resour. Conserv. Recycl. 2007, 49, 244–263.
28. Chang, Y.H.; Chang, N.B. Compatibility analysis of material and energy recovery in a regional solid waste management system. J. Air Waste Manag. Assoc. 2003, 53, 32–40.
29. Di Maria, F.; Micale, C. Impact of source segregation intensity of solid waste on fuel consumption and collection costs. Waste Manag. 2013, 33, 2170–2176.
30. Nguyen, T.T.T.; Wilson, B.G. Fuel consumption estimation for kerbside municipal solid waste (MSW) collection activities. Waste Manag. Res. 2010, 28, 289–297.
31. European Commission. Eurostat Database. Available online: http://epp.eurostat.ec.europa.eu/portal/page/portal/statistics/search_database (accessed on 1 December 2013).
32. Tchobanoglous, G.; Kreith, F. Handbook of Solid Waste Management; Mc-Graw-Hill: New York, NY, USA, 2002.
33. Murphy, J.D.; McKeogh, E. Technical, economic and environmental analysis of energy production from municipal solid waste. Renew. Energy 2004, 29, 1043–1057.
34. Rimaitytė, I.; Denafas, G.; Martuzevicius, D.; Kavaliauskas, A. Energy and environmental indicators of municipal solid waste incineration: Toward selection of an optimal waste management system. Pol. J. Environ. Stud. 2010, 19, 989–998.
35. Italian Institute of Statistics (ISTAT). Maps and Census Data. (in Italian). Available online: http://www.istat.it/it/strumenti/cartografia (accessed on 23 April 2013).
36. Institute for the Environmental Protection and Research—ISPRA. Rapporto rifiuti urbani (Municipal Waste Report); ISRRA: Rome, Italy, 2012.
37. Regione Basilicata. Piano regionale gestione rifiuti, Potenza, Italy. Available online: http://www.regione.basilicata.it/giunta/files/docs/DOCUMENT_FILE_242375.pdf (accessed on 20 April 2013).
38. Consonni, S.; Viganò, F. Material and energy recovery in integrated waste management systems: The potential for energy recovery. Waste Manag. 2011, 31, 2074–2084.
39. Valkenburg, C.; Gerber, M.A.; Walton, C.W.; Jones, S.B.; Thompson, B.L.; Stevens, D.J. Municipal Solid Waste (MSW) to Liquid Fuels Synthesis, Volume 1: Availability of Feedstock and Technology; U.S. Deptartment of Energy: Washington, DC, USA, 2008.
40. Cosmi, C.; Cuomo, V.; Macchiato, M.; Mangiamele, L.; Masi, S.; Salvia, M. Waste management modeling by MARKAL model: A case study for Basilicata Region. Environ. Model. Assess. 2000, 5, 19–27.
41. Cosmi, C.; Mancini, I.; Mangiamele, L.; Masi, S.; Salvia, M.; Macchiato, M. The management of urban waste at regional scale: The state of the art and its strategic evolution—Case study Basilicata Region (Southern Italy). Fresenius Environ. Bull. 2001, 10, 131–138.
42. Çengel, Y.A. Introduction to Thermodynamics and Heat Transfer; McGraw-Hill: New York, NY, USA, 2007.
43. Environmental Protection Agency—Ireland. Municipal Solid Waste—Pre-treatment & Residuals Management; IR-EPA: Wexford, Ireland, 2009.

44. Sue, D.C.; Chuang, C.C. Engineering design and energy analyses for combustion gas turbine based power generation system. Energy 2004, 29, 1183–1205.
45. Roher, A. Comparison of combined heat and power generation plants. ABB Review 1996, 3, 24–32.
46. Lombardi, F.; Lategano, E.; Cordiner, S.; Torretta, V. Waste incineration in rotary kilns: A new simulation combustion's tool to support design and technical change. Waste Manag. Res. 2013, 31, 739–750.

CHAPTER 9

Gaseous Emissions During Concurrent Combustion of Biomass and Non-Recyclable Municipal Solid Waste

RENÛ LARYEA-GOLDSMITH, JOHN OAKEY, AND NIGEL J. SIMMS

9.1 BACKGROUND

Concurrent combustion of biomass and municipal solid waste (MSW) offers a method of electricity and heat generation using renewable energy resources. At small scale, biomass is recognised as a form of renewable energy that is capable of meeting both heat and electricity demand most effectively in the form of combined heat and power, contributing towards international commitments to minimise environmental damage [1]. Further efficiency in using biomass can be obtained where thermal conversion occurs adjacent to areas of demand (such as cities) for cooling, i.e. "tri-generation," or combined cooling, heating and electrical power. Therefore, small-scale biomass combustion offers an excellent method to exploit heat energy. In contrast, wind turbines and large-scale pulverised fuel power stations are primarily used to produce electricity only, where

---

Gaseous Emissions during Concurrent Combustion of Biomass and Non-Recyclable Municipal Solid Waste. © Laryea-Goldsmith R, Oakey J, and Simms NJ. Chemistry Central Journal 5,4 (2011), hdoi:10.1186/1752-153X-5-4. The work is made available under the Creative Commons Attribution License, http://creativecommons.org/licenses/by/4.0/.

the pulverised fuel may contain biomass for co-firing. Within waste management government policy, energy recovery from MSW is seen as an essential requirement for diverting waste from landfill disposal. As a result of Europe-wide legislation and government targets to minimise the environmental impact of landfill disposal, for example it is forecast that energy recovery within UK is expected to comprise 25% of MSW disposal by 2020; recent rates are 10% [2]. Combustion of biomass is often assumed to provide nil net anthropogenic carbon dioxide emissions and carbon emissions associated with biomass collection and distribution activities are similar to those for fossil fuels [3] and therefore may be accepted as being of minimal damage to the environment. In contrast, the combustion of MSW is often considered to be detrimental both to human health and environmental management primarily because of various gaseous pollutants, most commonly public health risk concern about poly-chlorinated dibenzo-p-dioxins and dibenzofurans [4]. The most common assessment of biomass and MSW fuels has been to investigate combustion of biomass with coal, or coal with MSW. This paper is an investigation into the combustion of biomass and source separated MSW, specifically the extent of gaseous emissions.

Dried distillers' grains with solubles (DDGS) is investigated in this paper as an example biomass fuel, produced as an agricultural by-product in the manufacture of ethanol by fermentation. Historically, DDGS is used as animal feed but as a result of increased ethanol production, it is forecast to be of greater use as an energy source [5]. The MSW fuel was selected from the source-separated rejected material of a materials recycling facility. At least within the UK, combustion of MSW is considered by some to be a barrier against efforts to increase historically low levels of materials recycling. As a consequence, this material fraction was identified as a fuel source that could be exploited without impact upon the higher priority recycling of waste materials and therefore maintaining the "waste hierarchy" of sequentially preferred methods of waste management: a) reduction of waste b) reuse of waste materials c) recycling & composting d) energy recovery e) landfill with energy recovery f) landfill [2]. This material is termed in this paper: "materials recycling facility residue" (MRFR), to represent a truly residual waste fuel that does not conflict with materials recycling activities.

A pilot-scale fluidised bed combustor was used to burn biomass and MRFR in mixtures of various proportions. The effect of the MRFR waste fraction of the fuel mixture upon gaseous pollutant emissions during combustion was investigated using an FTIR flue gas analyser. This analytical technique suggested that addition of the MRFR to the fuel mixture inhibited emission of some pollutants. Use of rejected material from materials recycling facilities suggests that energy conversion of such waste is practicable and that variation in the composition of the fuel types can be tolerated, to represent the geographical constraints and seasonal changes in availability of these fuels.

TABLE 1: Summary of operation of fluidised bed combustor.

Fuel mixture	Fuel feed (kg h^{-1})	Inlet air (1 min $^{-1}$)	Combustion temperature (fluidised bed sand, °C)	Flue gas content CO_2 (% vol.)	Flue gas content O_2 (% vol.)	Flue gas content H_2O (% vol.)
DDGS	6	1400	855 (848)	8.5 (8.6)	11.2 (11.1)	9.0 (9.2)
DDGS-MRFR (90:10)	5	1400	888 (861)	8.3 (7.9)	11.6 (12.0)	8.5 (8.3)
DDGS-MRFR (80:20)	5	1400	864 (857)	8.8 (8.9)	11.5 (11.4)	9.4 (9.5)
DDGS-MRFR (50:50)	4	1400	791 (804)	6.9 (7.3)	13.0 (12.6)	7.0 (7.2)
DDGS-MRFR (30:70)	7	1400	818 (811)	10.2 (9.5)	9.7 (10.2)	9.4 (8.8)

Median values shown for temperature and flue gas data during periods of stable combustion. Corresponding values for entire duration of combustion of fuel mixtures are shown in parentheses. Duration of combustion tests were approximately three hours, except for DDGS 100% (two hours) and DDGS-MRFR 90:10 (four hours)

9.2 RESULTS

The general target control parameters for operation of the combustor were as follows: a) temperature 850°C (to mimic the mandatory minimum temperature required for a combustor consuming MSW as a fuel); b) flue gas oxygen content <10% (to indicate operational efficiency); c) duration at least two hours. The fluidised bed material consisted of silica sand and industrial limestone, 50% by weight of each material and total bulk mass 22 kg and the level of fluidisation was measured by use of a micromanometer (model MDG FC001, Furness Controls; Burgess UK), to monitor the differential pressure caused by the dynamic motion of gas through the fluidised bed particles. Operating conditions for the fluidised bed combustor are summarised in Table 1.

9.2.1 GAS POLLUTANT EMISSIONS

This paper is focussed on emissions of hydrogen chloride, nitrogen oxides and sulphur oxides from DDGS-MRFR combustion in the fluidised bed combustor. The following graphs (Figures 1, 2, 3, 4, 5, 6, 7, 8, 9 and 10) summarise these emissions for the range of fuel mixtures studied. Emissions data presented here correspond to periods during the combustion test when stable combustion conditions were maintained. The fuel mixtures of DDGS and MRFR consisted of the following DDGS fractions, by weight: 90%; 80%; 50%; 30%. Where graphs show temperature, this is a measure of the "in-bed" temperature of fluidised bed material within the fluidised bed combustor. The "box and whisker" graphs (Figures 1 and 2) are interpreted as follows: median value represented by the bold horizontal line; interquartile range (contains 50% of data set) represented by the area inside the box; horizontal lines above and below interquartile range box represent maximum and minimum values in data set; "outliers" represented by circle (○) symbol and defined as a datum 1.5 greater and/or lesser than the interquartile range.

9.3 DISCUSSION

Combustion of waste materials requires attention to ensure complete combustion is achieved in order to minimise gaseous pollutant emissions. The variation in temperature experienced during combustion of biomass and MRFR fuels may be attributed to volatile components in both fuel fractions being released in a random manner as the fuel mixes with the fluidised bed during combustion. The fuel was observed to ignite and burn often almost immediately upon entry into the freeboard area, as has been reported elsewhere during co-combustion with coal and paper, plastic waste [6]. Temperature per se has a significant impact upon reaction rates in general and in the case of combustion, gaseous emissions may either be enhanced or inhibited by increases in the combustion temperature. For example, carbon monoxide is an intermediate species and during combustion in the fluidised bed combustor is likely to be formed within localised regions of the fluidised bed material, where fuel-rich conditions may momentarily occur to prevent complete combustion to carbon dioxide because of insufficient oxygen being present in the local region. The rate of carbon monoxide oxidation is determined significantly by temperature; high combustion temperatures favour complete oxidation. In addition, the availability of free radicals is also a factor for carbon monoxide oxidation [7] and is a complex factor, affected by the presence of other pollutants such that competition for a limited concentration of free radicals will be determined by reaction kinetics.

To demonstrate the effect of free radicals, carbon monoxide emissions have been shown to be potentially higher in the presence of hydrogen chloride [8], an effect attributed to catalytic re-combination of free radicals that reduces the availability of free radical intermediates for consequent carbon monoxide oxidation. This observation is in itself affected by the combustion temperature; higher temperatures (which per se discourage carbon monoxide) reduces the impact of hydrogen chloride as a catalyst to prevent oxidation of carbon monoxide. This study has shown a less clear relationship between temperature, hydrogen chloride content and carbon monoxide emissions. Figure 5 shows that carbon monoxide emission was highest during combustion conditions of high temperature and highest hydrogen chloride content. However, during combustion of both pure DDGS (Figure 3) and DDGS-MRFR (50:50, Figure 4) a random pattern of emissions was found; highest carbon monoxide emissions occurred throughout various temperatures. Carbon monoxide emission was significantly higher for combustion of 30% DDGS fuel mixture and it is suggested that this is due to the high level of MRFR not having sufficient residence time within the combustion zone to achieve complete combustion of the fuel. These observations would be consistent with previous studies showing how at lower temperatures, the effect of hydrogen chloride is more significant in raising the carbon monoxide level [9]. Therefore, the result shows carbon monoxide emissions to be dependent not only on temperature, but is also affected by the chlorine content of the fuel. Although not measured, it is expected that other group 17 halides are also capable of similar effect if present in the fuel, as it has been reported similarly elsewhere that bromide and iodide cations are more effective inhibitors of carbon monoxide oxidation [10].

The sensitivity of carbon monoxide emission to the availability of free radicals has a consequence with respect to emissions of nitrogen and sulphur oxides. formation of these pollutants are themselves also determined by availability of free radicals [11] and it is claimed in general that sulphur reduction of free radicals such as of oxygen atoms reduces the reaction rates of thermal NOx formation. Therefore it is to be expected that competition for free radicals in order to achieve complete oxidation will be shown as related interactions between concentrations of oxides of carbon, nitrogen and sulphur. During combustion of 30% DDGS 70% MRFR, emissions of carbon monoxide occurred when nitric oxide and sulphur dioxide were at their highest (Figure 8). Use of additives in the fluidised bed has a direct impact on sulphur dioxide emissions and also indirect effects on other pollutants. Limestone was used in this study to inhibit agglomeration of the fluidised bed material with the fuel ash. Limestone reacts with sulphur in the fuel to prevent sulphur dioxide emission in the following series of reactions:

$$SO_2 + 12O_2 \rightarrow SO_3 \tag{1}$$

$$CaCO_3 \rightarrow CaO + CO_2 \tag{2}$$

$$CaO + SO_3 \rightarrow CaSO_4 \tag{3}$$

$$CaO + SO_2 + \tfrac{1}{2}O_2 \rightarrow CaSO_4 \qquad (4)$$

$$CaCO_3 + SO_2 + \tfrac{1}{2}O_2 \rightarrow CaSO_4 + CO_2 \qquad (5)$$

The absorption of sulphur dioxide is represented by the overall limestone sulphation reaction (equation 5), which is too slow to occur significantly within typical fluidised bed combustor conditions. Instead, sulphur dioxide absorption occurs more favourably via calcination of limestone (i.e. equation 2 followed by equation 4), whilst reaction with sulphur trioxide (equation 3) is catalysed by the presence of heavy metal salts most likely to originate from the MRFR fuel mixture component. In addition to sulphur control, the generation of lime (equation 2) has an impact on hydrogen chloride emissions. A secondary reason for the use of limestone in the fluidised bed is the expected removal of hydrogen chloride:

$$CaO + 2HCl \leftrightarrow CaCl_2 + H_2O \qquad (6)$$

This reaction between lime and hydrogen chloride is interesting in consideration of the melting point of calcium chloride (772°C) which occurs within various areas of the combustor such that calcium chloride deposits should be found within cooler zones of the combustor. For example, during combustion of DDGS-MRFR (90:10) the average temperature near the flue gas outlet zone of the fluidised bed combustor was found to be 671°C and therefore this suggestion appears plausible. However, competition between chloride and sulphur for reaction with limestone (equations 3, 4 and 6); reaction kinetics; and thermal stability of the calcium chloride deposits are additional factors to consider. The consequence of equation 6 is that limestone is not wholly effective in final removal of hydrogen chloride, but instead is involved in reactions causing both generation and consumption of hydrogen chloride. This conclusion is also made elsewhere [12] and would explain the detection of hydrogen chloride as shown in this paper.

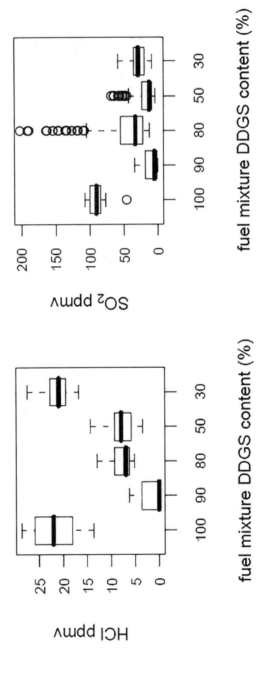

FIGURE 1: Summary of emissions of HCl and SO_2 for the fuel mixtures tested.

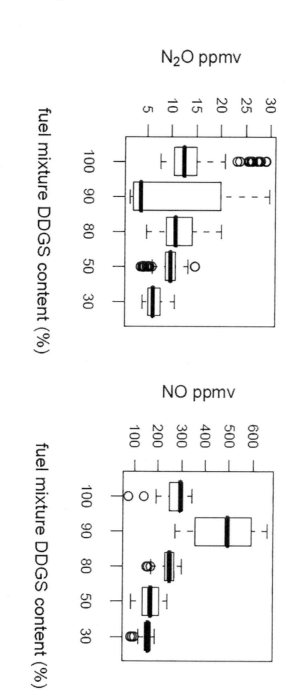

FIGURE 2: Summary of emission of nitrogen oxides for the fuel mixtures tested.

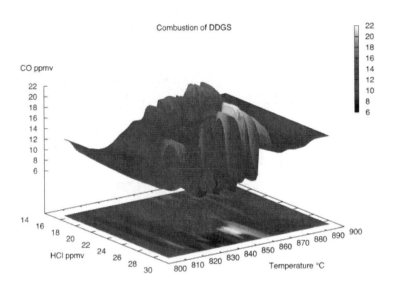

FIGURE 3: Relationship between simultaneous emissions of HCl and CO for a range of combustion temperatures; combustion of DDGS 100%.

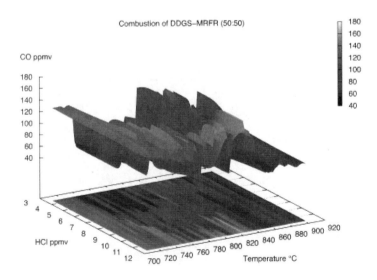

FIGURE 4: Relationship between simultaneous emissions of HCl and CO for a range of combustion temperatures; combustion of DDGS-MRFR 50:50.

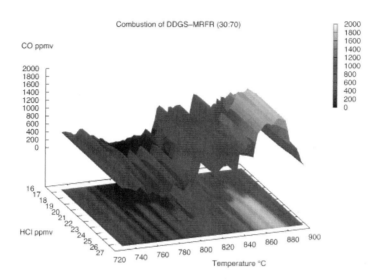

FIGURE 5: Relationship between simultaneous emissions of HCl and CO for a range of combustion temperatures; combustion of DDGS-MRFR 30:70.

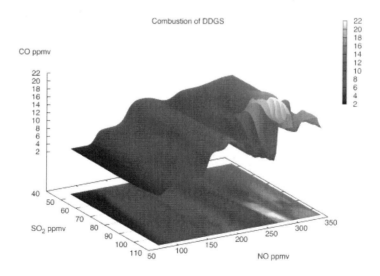

FIGURE 6: Relationship between simultaneous emissions of SO_2, NO and CO; combustion of DDGS 100%

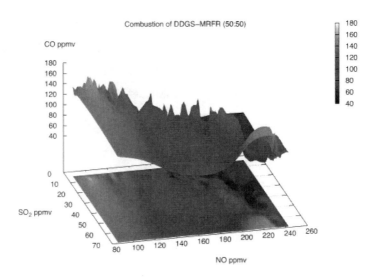

FIGURE 7: Relationship between simultaneous emissions of SO_2, NO and CO; combustion of DDGS-MRFR 50:50.

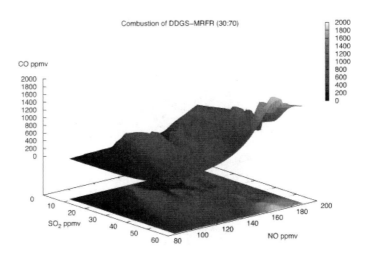

FIGURE 8: Relationship between simultaneous emissions of SO_2, NO and CO; combustion of DDGS-MRFR 30:70.

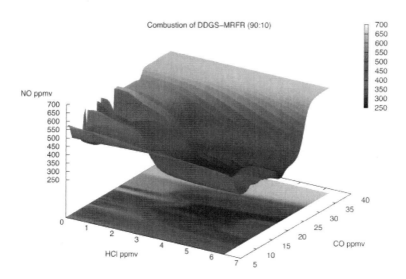

FIGURE 9: Relationship between simultaneous emissions of HCl, CO and NO; combustion of DDGS-MRFR 90:10.

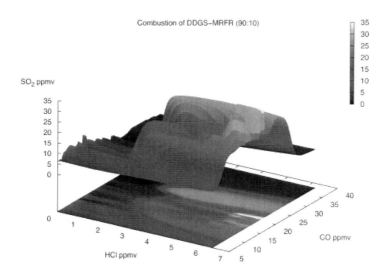

FIGURE 10: Relationship between simultaneous emissions of HCl, CO and SO_2; combustion of DDGS-MRFR 90:10.

Variations in the changes in emissions with the biomass fractions in the fuel mixtures are summarised in Figures 1 and 2. Of surprising interest are emissions of nitric oxide and sulphur dioxide, where the latter is reduced when the waste fraction is added to the biomass. For combustion of biomass fuels using a fluidised bed combustor, the source of nitric oxide emissions is attributed mostly to the nitrogen content in the fuel; this is in contrast to "thermal" nitric oxides emissions during coal combustion which is performed normally at higher temperatures. Nitrogen content is relatively high in DDGS due to the original protein content; laboratory analysis of the DDGS for these experiments was nearly 5% by weight (Table 2). This compares to typical nitrogen content for another biomass fuel such wheat, of 1.5% and is a cause for relatively higher NOx emissions with DDGS. In consideration of interactions between nitric oxide and sulphur dioxide as discussed earlier, competition for free radicals between these two pollutants appears to be most evident in the presence of hydrogen chloride and carbon monoxide (Figures 9 and 10). This kinetic competition, especially in the presence of hydrogen chloride would explain why sulphur dioxide emissions are suppressed. Of course, the primary treatment for sulphur dioxide is calcium carbonate, the mechanisms of which are described earlier (equations 1-5).

TABLE 2: Compositions of DDGS pellets, dimensions 6-8 mm diameter, 30 mm length.

	DDGS	MRFR
Carbon	45.2	39.8
Hydrogen	6.2	5.2
Oxygen	34.1	27.2
Nitrogen	4.91	0.9
Chlorine	0.2	0.3
Sulphur	0.56	0.1
Ash	4.1	18.4
Moisture	4.9	8.5
Calorific value (gross, MJ kg^{-1})	19.3	22

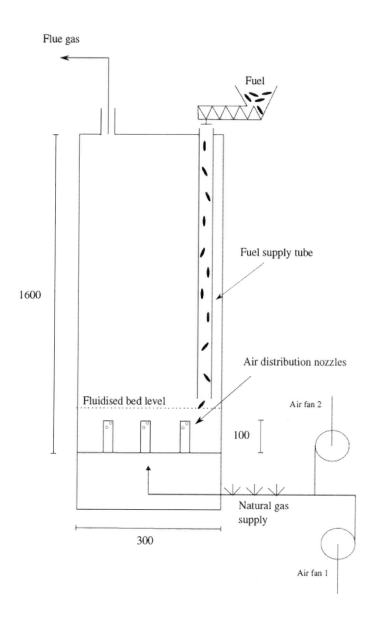

FIGURE 11: Fluidised bed combustor. Unit of measurement, millimetre.

9.4 CONCLUSIONS

Gas analysis of the combustion of biomass with non-recyclable waste suggests that stable combustion can be achieved and hence gaseous pollutant emissions can be minimised in conjunction with mandatory air pollution control equipment. There was a reduction observed in emissions of nitric oxide and sulphur dioxide when the biomass fraction is substituted for the waste fraction in combustion of the fuel mixtures. A likely explanation for this phenomenon is competition for free radical intermediates to prevent nitric oxide formation, at the expense of sulphur dioxide which itself is then partially converted to calcium sulphate by limestone. The oxidation of carbon monoxide is dependant significantly upon combustion temperature, but this effect may be overcome by the presence (as a catalyst) of hydrogen chloride which reduces the necessary availability of free radical intermediates.

Gas emission data sets from a range of fuel mixes and fluidised bed combustor operating conditions have been generated in this study. It is envisaged that these complex relationships can form the basis for future modelling activities. In summary, fluidised bed concurrent combustion is an appropriate technique to exploit biomass and municipal solid waste resources, without the use of fossil fuels. The addition of municipal solid waste—ideally the non-recyclable fraction as part of a sustainable waste management programme—to biomass combustion has the effect of reducing emissions of some gaseous pollutants.

9.5 EXPERIMENTAL

9.5.1 FLUIDISED BED COMBUSTOR DESIGN

The fluidised bed combustor (Figure 11) is based upon the bubbling fluidised bed design, in which hot air of sufficient velocity (in this apparatus, approximately 0.7 m s^{-1} for inlet air at 500°C) is injected into a bed of inert particles (e.g. sand) to cause the particle to move dynamically in similar way to a viscous liquid. The following diagram shows the features of the combustor used for this experiment. The combustion zone is of cuboid shape, the sides being 300 mm length, height approximately 1600 mm. The fluidisation air distribution grid comprises nine nozzles, each nozzle having 12 holes. The nozzles are arranged into three linear rows. The fuel feed is of Archimedes screw type, primarily tolerable of pelletised fuels and fuel chips less than 40 mm particle size. Gas analysis was performed using an FTIR flue gas analyser (Protir 204 M, Protea, Crewe UK), where the flue gas sample was extracted via a sample point maintained at a temperature of 180°C and at a location approximately 4-5 metres from the outlet of the combustor.

9.5.2 FLUIDISED BED MATERIAL

Silica sand (97% quartz, remainder tridymite, crystobalite) of size specification 0.5-1 mm was used (Garside Sands; Leighton Buzzard, UK). Limestone granules (5 mm particle size, 39.4% calcium content) were obtained from Tarmac (Buxton, UK).

9.5.3 FUEL CHARACTERISTICS

The bulk composition of the MRFR fuel fraction selected for this experiment was predominantly 90% plastics 10% fibres (paper, cardboard, wood). The material was comminuted to an average particle size of 20 mm using a laboratory scale cutting mill (Retsch; Haan, Germany). All values shown on "as received" weight percentages basis.

REFERENCES

1. DEFRA: UK Biomass Strategy. In Tech rep. Department for Food, Energy and Rural Affairs, London; 2007.
2. DEFRA: Waste Strategy for England 2007. In White paper. Department of Food and Rural Affairs, London; 2007.
3. Carbon Trust: Biomass sector review for the Carbon Trust. In Tech rep. Carbon Trust, London; 2005.

4. Hester R, Harrison R, (Eds): Waste Incineration and the Environment. No. 2 in Issues in Environmental Science and Technology. Cambridge: Royal Society of Chemistry; 1994.
5. Tiffany DG, Morey RV, Kam MD: Transition to a Bioeconomy: Integration of Agricultural and Energy Systems. In Use of distillers by-products and corn stover as fuels for ethanol plants. Atlanta: Farm Foundation; 2008.
6. Boavida D, Abelha P, Gulyurtlu I, Cabrita I: Co-combustion of coal and non-recyclable paper and plastic waste in a fluidised bed reactor. Fuel 2003, 82(15-17):1931-1938.
7. van Loo S, Koppejan J, (Eds): The handbook of biomass combustion and co-firing. London: Earthscan; 2008.
8. Wei X, Schnell U, Han X, Hein KRG: Interactions of CO, HCl, and SOx in pulverised coal flames. Fuel 2004, 83(9):1227-1233.
9. Gokulakrishnan P, Lawrence A: An experimental study of the inhibiting effect of chlorine in a fluidized bed combustor. Combustion and Flame 1999, 116(4):640-652.
10. Julien S, Brereton CMH, Lim CJ, Grace JR, Anthony EJ: The effect of halides on emissions from circulating fluidized bed combustion of fossil fuels. Fuel 1996, 75(14):1655-1663.
11. Glassman I, Yetter RA: Combustion. Fourth edition. London: Academic Press Inc. (London) Ltd; 2008.
12. Lawrence AD, Bu J: The reactions between Ca-based solids and gases representative of those found in a fluidized-bed incinerator. Chemical Engineering Science 2000, 55(24):6129-6137.

CHAPTER 10

Environmental Effects of Sewage Sludge Carbonization and Other Treatment Alternatives

NING-YI WANG, CHUN-HAO SHIH, PEI-TE CHIUEH, AND YU-FONG HUANG

10.1 INTRODUCTION

Because of high population growth and urban planning in Taiwan, the prevalence of public sewage systems reached 30% of the population in 2012, and is expected to increase to 36% by 2014 [1]. Thus, sewage sludge production will increase with the expansion of the sewage treatment system, and should reach up to 1040 t/day by 2014 [2]. Sewage sludge generally contains pollutants such as human pathogenic organisms, and must be disposed of in ways that reduce environmental and public health effects.

Most sewage sludge in Taiwan is currently disposed of in landfills, with the remainder being co-incinerated with municipal solid waste (MSW). Existing crane and grapple-feeding devices have difficulty handling the

---

Environmental Effects of Sewage Sludge Carbonization and Other Treatment Alternatives. © Wang N-Y, Shih C-H, Chiueh P-T, and Huang Y-F. Energies **6,**2 (2013). doi:10.3390/en6020871. Licensed under a Creative Commons Attribution 3.0 Unported License, http://creativecommons.org/licenses/by/3.0/.

pasty sludge cake with MSW, and the sludge degrades combustion efficiency. Thus, the co-incineration ratio is limited. Some MSW incineration plants even ban sewage sludge. The scarcity of available landfills and limited capacity of co-incineration are pressing problems. Other solutions for handling sewage sludge in more environmentally friendly ways, and recovering its energy, have caused great concern in recent years.

The imported energy ratio in Taiwan is as high as 99.4%, and energy security is unfavorable [3]. Finding alternative energy sources, such as bioenergy, is necessary. Carbonization technology can transform sludge into a carbon-containing product that can be used as biocoal and co-fired with fossil coal to generate electricity in power plants [4–8]. Carbonizing sludge reduces its volume to approximately one-eighth of the sludge cake, increases its calorific value, removes its odor, and improves its combustibility and grindability, making it a better co-firing material for pulverized coal power plants [4,5,9]. Reference plants applying sewage sludge carbonization technology in Japan and North America have successfully demonstrated its feasibility [8–11]. These applications advance the goals of using sewage sludge as an energy resource and simultaneously reducing greenhouse gas emissions and coal extraction.

Life cycle assessment (LCA) is a method of evaluating the environmental effects associated with a product, process, or service throughout its life cycle. The LCA method is generally performed according to ISO14040 standards, which define the principles and framework of LCA [12]. Researchers have also applied LCA to sewage sludge management. Hospido et al. [13] compared three alternative sewage sludge post-treatments (agricultural use, incineration, and pyrolysis) and then assessed the energy reuse strategy used in pyrolysis. Hong et al. [14] combined LCA and LCC (i.e., life cycle cost) to estimate the environmental and economic effects of six alternative sewage sludge treatments. Their results indicate that dewatered sludge combined with electric melting is an environmentally optimal and economically affordable method. Murray et al. [15] also applied life cycle environmental and life cycle cost assessments to nine alternative sewage sludge treatments. Their results indicate that coal-fired incineration is the most environmentally and economically costly of all treatments. However, no study has presented the LCA of sewage sludge carbonization. Because carbonization is increasingly being adopted in several countries and is a candidate for sludge treatment in Taiwan, understanding the potential effects of this biomass usage method is necessary.

The objectives of this study are to simulate the sewage sludge carbonization process, using local sludge properties, to evaluate the environmental effects and benefits of the carbonization process using LCA. This study also uses LCA to investigate current approaches for sewage sludge treatment, including direct landfills, co-incineration with MSW, and mono-incineration for comparison.

10.2 SIMULATION OF THE SEWAGE SLUDGE CARBONIZATION PROCESS

Because no inventory is available within the existing LCA database applicable to carbonization, this study adopts an energy model developed by Maski et al. for biomass pretreatment [16] and previous research results of biomass torrefaction [17–19]. This study also refers to a batch-type carbonation experiment, conducted by Park and Jang [4], specific to dried sewage sludge at 300–500 °C for 30 min. Koga et al. [8] also reported a sewage sludge carbonization system handling 40–60 kg/h of dewatered sludge at a 500 °C carbonation temperature to produce biocoal. Therefore, the simulated carbonization process in this study assumed dewatered sludge to be bone dried at 100 °C, and subsequently carbonized at 450 °C for 30 min in the absence of oxygen. Carbonized liquid and volatile gases were collected and recovered for their heat energy [20,21], which was supplied to the drying and carbonization units through a combustor and heat exchanger. The final carbonized product or biocoal, which had properties similar to fossil coal, can generate carbon neutral bioenergy at a pulverized coal power plant.

The following equations were used to simulate the sewage sludge carbonization process:

1. The energy use of a drying unit ($E_{R,D}$, MJ/kg) is represented by:

$$E_{R,D} = \left\{ \left(\frac{M_{wet}}{100}\right) \times \left[C_{p,w} \times (373 - T_i) + L_{v,w}\right] + \left(\frac{DB_{wet}}{100}\right) \times C_{p,b} \times (373 - T_i) \right\} \times e_{f,d}^{-1} \quad (1)$$

where M_{wet} (wt %) is the moisture content of sewage sludge; DB_{wet} (wt %) is the percentage of dry solid in sewage sludge (note: $M_{wet} + DB_{wet} = 100$ wt %); $C_{p,w}$ (MJ/kg K) is the specific heat of water = 0.004187 MJ/kg K; T_i (K) is the initial temperature of sewage sludge = 298 K; $L_{v,w}$ (MJ/kg) is the latent heat of water at its boiling point = 2.27 MJ/kg; $C_{p,b}$ (MJ/kg K) is the specific heat of sewage sludge = 0.001763 MJ/kg K [18]; and $e_{f,D}$ is the efficiency of the drying unit (assumed to be 0.85 in this study, a relatively high efficiency).

2. The energy use of a carbonization unit ($E_{R,C}$, MJ/kg) is represented by:

$$E_{R,C} = \frac{C_{p,b} \times (T_C - 373)}{e_{f,c}} \tag{2}$$

where T_C (K) is the carbonization temperature, and $e_{f,C}$ is the efficiency of the carbonization unit, set to 0.85 in this study.

3. In a combustor and heat exchanger, available energy is derived from the combustion of volatile gas and carbonized liquid, where available energy from volatile gas ($E_{A,CG}$, MJ/kg) and available energy from carbonized liquid ($E_{A,CL}$, MJ/kg) are represented by:

$$E_{A,CG} = \frac{1}{3} LHV_{volatile} \times DB \times y_{MG} \times e_{fmc} \times (1 - H_L) \tag{3}$$

and:

$$E_{A,CL} = \frac{1}{3} LHV_{liquid} \times DB \times y_{ML} \times e_{f,c} \times (1 - H_L) \tag{4}$$

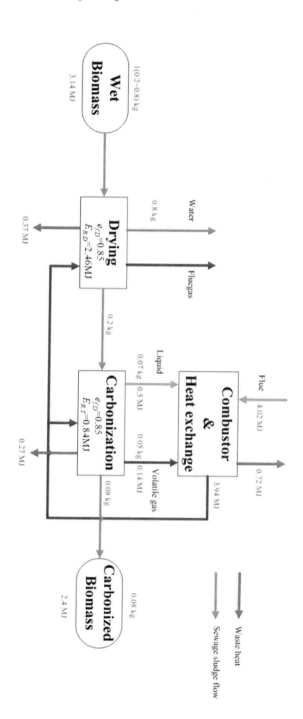

FIGURE 1: Mass and energy balances of the sewage sludge carbonization process.

The terms $LHV_{volatile}$ (MJ/kg) and LHV_{liquid} (MJ/kg) represent the heating value of the volatile gas and carbonized liquid generated by the carbonization unit, respectively, and D_B (kg) is the weight of dried sludge, y_{MG} is the volatile gas yield, y_{ML} is the carbonized liquid yield [calculated by Equation (5)], $e_{f,C}$ is the efficiency of the combustor unit (assumed to be 0.85), and HL is the heat loss of the heat exchanger (assumed to be 0.5%).

4. The product yield (y_M) is defined according to mass by [22]:

$$y_M = \left(\frac{m_{out}}{m_{in}}\right)_{daf} \quad (5)$$

where m_{in} is the mass of the biomass input, and m_{out} is the mass of product output of a carbonization unit (note: daf = dry and ash free).

TABLE 1: Sewage sludge characteristics.

Dewatered sludge characteristics*	Value
Moisture (wt %)	80
High heating value dry (MJ/kg)	15.18
Proximate analysis (dry basis, wt %)	
Ash content	35.2
Volatile matter	64.8
Elemental analysis (dry and ash free basis, wt %)	
C	54.60
H	7.69
N	4.52
O	30.29
S	2.52
Experimental results of carbonization yield (dry and ash free basis, wt %)**	
Solid yield	39.14
Liquid yield	34.09
Volatile gas	26.77

Notes: * Source: [2]; ** Source: [23,24], recalculated by this study at 450 °C.

Figure 1 shows the simulation results of sewage sludge carbonization based on the properties of dewatered sewage sludge after anaerobic digestion from a local sewage treatment plant and experimental results of carbonization yield [23,24] (Table 1). Mass and energy balance calculations show a 0.08 (daf) product yield (yM), 3.19 MJ/kg total energy required of units, 0.14 MJ/kg available energy from volatile gas ($E_{A,CG}$), and 0.5 MJ/kg available energy from carbonized liquid ($E_{A,CL}$). The supplementary information (Table A1) explains the nomenclature and provides parameter values. Although the simulation process is a simplified form, this approach presents a feasible method to access the LCA of sewage sludge carbonization. By changing the efficiencies of the drying unit and carbonization unit from 0.85 to 0.65 (decreasing 20%), the overall required energy increases 25% and the available energy from volatile gas and carbonized liquid decreases 19%. The influence of the assumed efficiency of each unit on the LCA results can be estimated accordingly.

10.3 MATERIALS AND METHODS OF LCA

The LCA software SimaPro 7.2 (PRé Consultants, Amersfoort, The Netherlands) was used to assess the environmental effects of four sewage sludge treatment scenarios: carbonization, mono-incineration, landfill, and co-incineration with MSW. The IMPACT2002+ model (included in SimaPro 7.2) was used to characterize environmental effects [25] by combining midpoint assessments with various damage categories.

10.3.1 FUNCTIONAL UNIT

The functional unit is defined as the unit for comparison in a life cycle inventory. Specifically, this study adopts the management of 1 t of dewatered sludge (moisture content 80%) as the functional unit on which all material and energy use, energy recovery, and emissions are based.

196 Solid Waste as a Renewable Resource: Methodologies

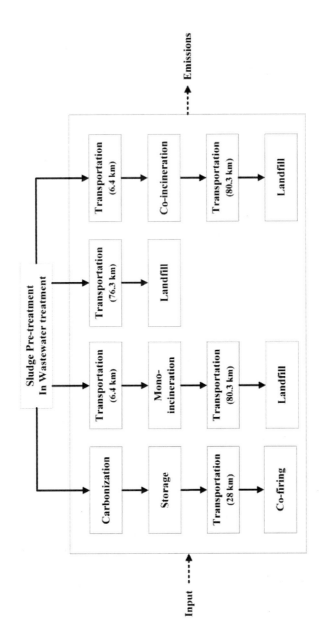

FIGURE 2: System boundary. Values in parentheses are transportation distances.

10.3.2 SYSTEM BOUNDARIES

Dewatered sludge after anaerobic digestion from the Dihua sewage treatment plant in Taipei City (Taiwan) was adopted as an example for evaluation in this study, and its characteristics are detailed in reference [26]. Figure 2 shows the system boundaries of the study: carbonization and mono-incineration were hypothesized alternatives, whereas landfill and co-incineration with MSW were modeled according to current practice. The considerations for each system included the following factors:

1. Carbonization of dewatered sludge and co-firing of biocoal (1% by heat input) at the Linkou coal power plant, New Taipei City, Taiwan, considering coal substitution, electricity generation, and heat recovery.
2. Mono-incineration of dewatered sludge at certain industrial waste incineration plants without production of electrical energy, considering heat recovery.
3. Sanitary landfill of dewatered sludge at Wujie Township, Yilan County, Taiwan, without considering methane recovery.
4. Co-incineration of dewatered sludge (3% by weight) with MSW at the Beitou MSW incineration plant, New Taipei City, Taiwan, in which waste to electricity was considered.

10.3.3 INVENTORY DATA SOURCE

10.3.3.1 ENERGY USE AND RECOVERY

This study investigates the inventories of energy use and recovery to assess the potential net energy benefits of different sludge treatments. For all scenarios, electricity was the primary energy consumed, and its impacts on environment were characterized by the actual electricity structure in Taiwan (coal-fired: 50%; LNG-fired: 25%; nuclear: 17%; oil-fired: 4%; hydro: 3%; waste: 1%). For the assessment of energy recovery in the carbonization scenario, the quantity of recycled heat (640 MJ), coal substitution (80 kg), and electricity generation (105 kWh) resulting from co-firing was derived from simulation results and heating values conversion of biocoal (Figure 1). The inventory also includes the effects of sludge transportation by trucks (>28 t) and coal substitution on coal mining and ocean shipping.

Table A2 lists energy and material inputs for mono-incineration scenario. The mono-incineration scenario recovered heat energy (349.4 MJ) during the incineration process [13]. The co-incineration scenario generated electricity (0.21 kWh) by the co-incineration of sludge with MSW [27]. The landfill scenario recovered no energy (methane).

10.3.3.2 DATA QUALITY

Other inventory data used in this study were obtained from three main sources:

1. SimaPro 7.2 databases: The LCA, Ecoinvent, Industry data 2.0, IDEMAT 2001, and LCAfood databases contained in SimaPro software were used for the inventory of the four scenarios (carbonization, direct landfills, co-incineration with municipal solid waste, and mono-incineration). This study also investigates the variables of input, output, emission, and waste disposal stages.
2. Operational data: Data used for the co-incineration scenario, such as electricity generation and electricity consumption, were obtained from practical data logging at the Beitou MSW incinerator Taiwan.
3. Literature and theoretical calculation: Neither carbonization nor mono-incineration of sewage sludge are practiced in Taiwan. Therefore, the inventory of the mono-incineration scenario was derived from previous studies, and theoretical calculations were applied to the carbonization scenario based on Equations (1)–(5).

TABLE 2: Characterization of the carbonization scenario.

Impact category	Unit	Carbonization and co-firing Total	Carbonization process					Co-firing of biocoal and coal in power plant		
			Drying	Carbonization	Energy Reuse	Carbonization Facility	Biocoal Storage	Co-firing	Alternative coal	Electricity production
Carcinogens	kg C2H$_3$Cl eq.	1.98	1.66	0.51	0	0	0	0.32	−0.08	−0.11
Non-carcinogens	kg C$_2$H$_3$Cl eq.	2.82	0.23	0.04	−0.02	0	0	2.96	−0.17	−0.22
Respiratory inorganics	kg PM$_{2.5}$ eq.	0.08	0.05	0.01	0.00	0	0.04	0.08	−0.04	−0.06
Ionizing radiation	Bq C^{-14} eq.	−1182	1556.3	158.23	−39.90	0.21	0	22.68	−274.61	−2,605.29
Ozone layer depletion	kg CFC^{-11} eq.	0.00003	0.00	0.00	0.00	0	0	0.000001	−0.000001	−0.000002
Respiratory organics	kg C$_2$H$_4$ eq.	0.03	0.05	0.01	−0.01	0	0	0.004	−0.02	−0.01
Aquatic ecotoxicity	kg TEG water	−22,412	3,952.9	671.52	−326.1	0.94	0	1,546	−23,768.1	−4,489.73
Terrestrial ecotoxicity	kg TEG soil	−6,003	933.3	161.68	−79.61	0.31	0	60.87	−6,044.6	−1,035.48
Terrestrial acid/nutria	kg SO$_2$ eq.	1.12	1.26	0.24	−0.12	0	0	2.47	−1.45	−1.29
Land occupation	m^2 land arable · year	−0.85	0.15	0.02	−0.01	0	0.06	0.02	−0.88	−0.22
Aquatic acidification	kg SO$_2$ eq.	0.36	0.34	0.06	−0.03	0	0	0.65	−0.27	−0.39

TABLE 2: *Cont.*

Impact category	Unit	Carbonization and co-firing	Carbonization process					Co-firing of biocoal and coal in power plant		
		Total	Drying	Carbonization	Energy Reuse	Carbonization Facility	Biocoal Storage	Co-firing	Alternative coal	Electricity production
Aquatic eutrophication	kg PO_4 eq.	-0.0004	0.00	0.00	0.00	0	0.0000002	0.0001	-0.0006	-0.001
Global warming	kg CO_2 eq.	146.62	209.02	55.79	-34.15	0.01	0	9.37	-16.00	-77.42
Non-renewable energy	MJ primary	-1122.9	3,896.4	1,062.72	-655.6	0.15	0	15.33	-4,178.8	-1,263.00
Mineral extraction	MJ surplus	-0.02	0.18	0.04	-0.02	0	0	0.02	-0.13	-0.12

10.4 IMPACT ASSESSMENT AND DISCUSSION

10.4.1 CARBONIZATION SCENARIO

The sewage sludge carbonization process was divided into several steps to identify the major sources of environmental effects and their consequences (Table 2). Results indicate that the drying unit created the highest environmental effect of all categories. Non-renewable energy and global warming were the categories most affected by drying and carbonization because of their high energy use. Recycling heat during carbonization had a positive effect on all categories. In addition, biocoal storage increased particulate emissions and land use.

This study assumes that the biocoal produced from the carbonization process will be used for co-firing with coal in power plants. As Table 2 shows, the overall results of the carbonization scenario generated positive effects in the categories of terrestrial ecotoxicity, aquatic ecotoxicity, land occupation, ionizing radiation, aquatic eutrophication, non-renewable energy, and mineral extraction. These benefits resulted primarily from coal substitution and bioelectricity generation.

10.4.2 COMPARISON OF SCENARIOS

In addition to the benefits and advantages of the carbonization scenario, this study presents a comparison of the results of four sewage sludge treatment scenarios. Figure 3 shows the characterization of the midpoint environmental effects for these scenarios. The carbonization scenario had the highest effect on ozone layer depletion and respiratory organics. Mono-incineration had the greatest effect on mineral extraction, non-renewable energy, aquatic acidification, terrestrial acid/nutrients, terrestrial ecotoxicity, ionizing radiation, and respiratory inorganics. The landfill scenario had the greatest effect on global warming, aquatic eutrophication, and land occupation. Co-incineration had the greatest effect on carcinogens and non-carcinogens and aquatic ecotoxicity.

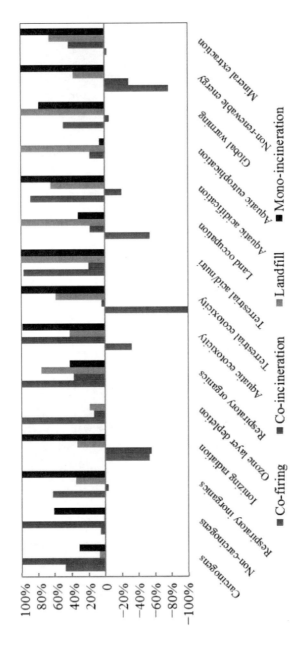

FIGURE 3: Characterization of the mid-point environmental effects of four sludge-handling scenarios.

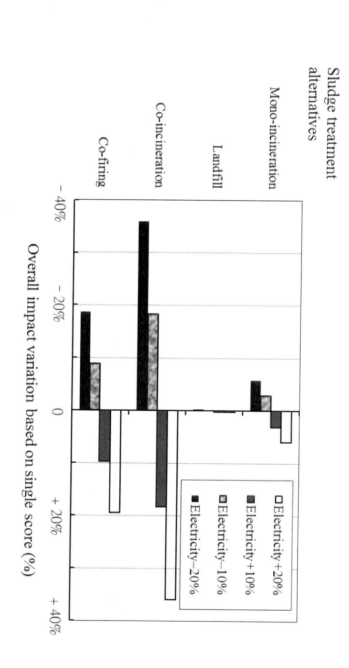

FIGURE 4: Sensitivity analysis of electricity consumption for four sludge-handling scenarios.

TABLE 3: Results of normalization, single score, and GHG emission in four sludge-handling scenarios

Item	Normalized results of damage categories				Single score	GHG emission
	Human health	Ecosystem quality	Climate change	Resources		
Unit	-	-	-	-	Pt	kg CO_2 eq.
Carbonization	0.0095	-0.0035	0.0148	-0.0074	0.013428	146.6
Co-incineration	0.0204	0.0003	-0.0016	-0.0028	0.016414	-15.4
Landfill	0.0048	0.0008	0.0300	0.0036	0.039208	296.9
Mono-incineration	0.0246	0.0012	0.0236	0.0097	0.059045	233.2

Table 3 shows the normalized results of the four scenarios for various damage categories: human health, ecosystem quality, climate change, and resources. Regarding damage to human health, mono-incineration and co-incineration damaged human health than the other scenarios. The effects of mono-incineration were caused primarily by the emission of particulates and nitrogen oxides, combined with the effluence of antimony and arsenic ions. The carbonization scenario had a beneficial effect on ecosystem quality because of the substitution of biocoal for some coal in the co-firing process. The landfill scenario had the greatest effect on climate change because of the greenhouse gas emitted during the landfill process. Mono-incineration had the greatest negative effect on resources because of its high energy use. Carbonization and co-incineration had positive effects on resources because of coal substitution and bioelectricity generation.

The single score column in Table 3 shows that the overall degree of environmental effect was mono-incineration > landfill > co-incineration > carbonization. Thus, carbonization combined with co-firing was the best scenario because it had the lowest environmental effect, followed by co-incineration, landfill, and mono-incineration in descending order. Table 3 shows the potential effect of the scenarios on global warming. Ranked in descending order of their greenhouse gas emissions, the scenarios were landfill (296.9 kg CO_2 eq.) > mono-incineration (232.2 kg CO_2 eq.) > carbonization (146.1 kg CO_2 eq.) > co-incineration (−15.4 kg CO_2 eq.). Landfills had the highest value because of the methane and carbon dioxide released during sludge decomposition. The greenhouse gas emissions of mono-incineration and carbonization primarily resulted from energy use. Only co-incineration showed a reduction in greenhouse gas emissions because of the bioelectricity generated during the incineration process.

The evaluation of greenhouse gas emissions in this study is similar to that of other studies, with the exception of the novel presented carbonization scenario. Houillon and Jolliet [28] assessed a sludge treatment method consisting of landfill, incineration, and pyrolysis. Their results show that the landfill treatment returned the highest greenhouse gas emissions. Lundin et al. [29] and Svanstrom et al. [30] compared greenhouse gas emissions of sludge treatments, indicating that co-incineration was beneficial in reducing greenhouse gas emissions. However, the greenhouse gas emission calculations for the carbonization scenario were higher than that for co-incineration because of the energy used (in particular by the drying unit) during the carbonization process. According to the current study results, carbonization combined with co-firing is less advantageous than co-incineration with MSW alone because of two possible reasons. In the current study, the water content of the dried sludge was assumed to be 0%, whereas Koga et al. [8] reported that dewatered sludge was only partially dried (i.e., until reaching 25% moisture content) before the carbonization step. In addition, this study assumes that the carbonization temperature is 450 °C (compared with 500 °C by Koga et al. [8]). This lower temperature may be more efficient and reduce energy use [4]. However, Lundin et al. [29] showed that the cost of co-incineration with waste was higher than using it in agricultural applications and incineration or fractionation combined with phosphorus recovery. Therefore, the co-incineration scenario may incur higher costs than other scenarios. Thus, environmental and economic effects should be cautiously considered when evaluating potential sludge treatments in the future.

10.4.3 SENSITIVITY ANALYSIS

Energy use is crucial in sewage sludge treatment options. Because all four scenarios in this study consume electricity, sensitivity analysis was performed using electricity consumption variances of ±10% and ±20%.

The variations of single scores of the four sludge treatment scenarios were analyzed accordingly (Table 4). Results indicate that the co-incineration scenario was most sensitive to variation in electricity consumption. When the electricity consumption increased by 20%, the overall effect increased by 36% (Figure 4). Variation of electricity consumption had little effect on the environmental impacts of the landfill scenario.

TABLE 4: Results of single scores for electricity consumption variances.

Variation	Co-firing	Co-incineration	Landfill	Mono-incineration
Electricity −20%	1.09×10^{-2}	1.05×10^{-2}	3.91×10^{-2}	5.56×10^{-2}
Electricity −10%	1.22×10^{-2}	1.34×10^{-2}	3.92×10^{-2}	5.73×10^{-2}
Original case	1.34×10^{-2}	1.64×10^{-2}	3.92×10^{-2}	5.90×10^{-2}
Electricity +10%	1.47×10^{-2}	1.94×10^{-2}	3.93×10^{-2}	6.08×10^{-2}
Electricity +20%	1.60×10^{-2}	2.23×10^{-2}	3.93×10^{-2}	6.25×10^{-2}

10.5 CONCLUSIONS

This study presents an assessment of the environmental effects of four sewage sludge treatment options: carbonization, mono-incineration, direct landfills, and co-incineration with municipal solid waste. This study uses an energy model to simulate the process of sewage sludge carbonization and produces theoretical energy and mass balance data for conducting the LCA. The results of the four treatment scenarios show that carbonization was the most preferable sludge-handling option overall, followed by co-incineration, landfills, and mono-incineration in descending order.

However, the co-incineration option emitted less greenhouse gases than carbonization because the overall energy recovery ratio of electricity was higher during the incineration process than during carbonization. Although this analysis considers heat recovery during carbonization, electricity generation, and coal substitution during co-firing, the energy used in drying the dewatered sludge emitted more greenhouse gases, contributing greatly to the damage category of climate change. However, changing both the feeding water content after the drying process and the carbonization temperature may mitigate the energy use of the carbonization scenario.

The aspect of cost must also be considered in the assessment and selection of sewage sludge treatment options. Because the application of sewage sludge carbonization is currently receiving great attention from municipal authorities, the significance of sewage sludge as a valuable energy source may increase even more.

REFERENCES

1. Construction and Planning Agency of Taiwan Home Page. Available online: http://www.cpami.gov.tw/chinese/index.php?option=com_content&view=article&id=13545&Itemid=13123 (accessed on 7 January 2012).
2. Assessment and Planning of Sewage Sludge Disposal; Construction and Planning Agency: Taipei, Taiwan, 2011.
3. Statistics and Analysis of CO2 Emissions from Fuels Combustion; Bureau of Energy: Taipei, Taiwan, 2011.
4. Park, S.W.; Jang, C.H. Characteristics of carbonized sludge for co-combustion in pulverized coal power plants. Waste Manag. 2011, 31, 523–529.
5. Kim, Y.J.; Choi, J.H.; Kim, J.H.; Lee, C.S.; Qureshi, T.I. Production and effective utilization of carbonized sludge of industrial wastewater treatment plant. J. Chem. Soc. Pak. 2010, 32, 7–12.
6. Oda, T. Making Fuel Charcoal from Sewage Sludge for Thermal Power Generation Plant—First in Japan. In Proceedings of Water Environment Federation Technical Exhibition & Conference, San Diego, California, USA, 13–17 October 2007.
7. Park, S.W.; Jang, C.H. Effects of carbonization and solvent-extraction on change in fuel characteristics of sewage sludge. Bioresour. Technol. 2011, 102, 8205–8210.
8. Koga, Y.; Endo, Y.; Oonuki, H.; Kakurata, K.; Amari, T.; Ose, K. Biomass solid production from sewage sludge with pyrolysis and co-firing in coal power plant fuel. Mitsubishi Heavy Ind. Tech. Rev. 2007, 44, 44–48.
9. Kimuro, Y.; Furubayashi, T.; Nakata, T. An inventory analysis of sewage energy system [in Japanese]. J. Jpn. Inst. Energy 2011, 90, 247–257.
10. Koga, Y.; Mizutani, H.; Tsuneizumi, S.; Yamamoto, H.; Tabata, M.; Amari, T. New biomass utilization technologies such as methane fermentation and pyrolysis. Mitsubishi Heavy Ind. Tech. Rev. 2007, 44, 39–43.
11. Bolin, K.M.; Dooley, B.; Kearney, R.J. Carbonization Technology Converts Biosolids to an Economical, Renewable Fuel. In Proceedings of Moving Forward Wastewater Biosolids Sustainability: Technical, Managerial, and Public Synergy, New Brunswick, Canada, 24–27 January 2007; pp. 591–597.
12. ISO 14040:2006 Environmental Management—Life Cycle Assessment—Principles and Framework; International Organization for Standardization (ISO): Geneva, Switzerland, 2006.

13. Hospido, A.; Moreira, M.T.; Martin, M.; Rigola, M.; Feijoo, G. Environmental evaluation of different treatment processes for sludge from urban wastewater treatments: Anaerobic digestion versus thermal processes. Int. J. Life Cycle Assess. 2005, 10, 336–345.
14. Hong, J.L.; Hong, J.M.; Otaki, M.; Jolliet, O. Environmental and economic life cycle assessment for sewage sludge treatment processes in Japan. Waste Manag. 2009, 29, 696–703.
15. Murray, A.; Horvath, A.; Nelson, K.L. Hybrid life-cycle environmental and cost inventory of sewage sludge treatment and end-use scenarios: A case study from China. Environ. Sci. Technol. 2008, 42, 3163–3169.
16. Maski, D.; Darr, M.; Anex, R. Torrefaction of Cellulosic Biomass Upgrading—Energy and Cost Model. In Proceedings of American Society of Agricultural and Biological Engineers Annual International Meeting, Pittsburgh, PA, USA, 20–23 June 2010; pp. 4443–4460.
17. Syu, F.S.; Chiueh, P.T. Process simulation of rice straw torrefaction. Sustain. Environ. Res. 2012, 22, 177–183.
18. Arlabosse, P.; Chavez, S.; Prevot, C. Drying of municipal sewage sludge: From a laboratory scale batch indirect dryer to the paddle dryer. Braz. J. Chem. Eng. 2005, 22, 227–232.
19. Chiueh, P.T.; Lee, K.C.; Syu, F.S.; Lo, S.L. Implications of biomass pretreatment to cost and carbon emissions: Case study of rice straw and Pennisetum in Taiwan. Bioresour. Technol. 2012, 108, 285–294.
20. Chun, Y.N.; Kim, S.C.; Yoshikawa, K. Pyrolysis gasification of dried sewage sludge in a combined screw and rotary kiln gasifier. Appl. Energy 2011, 88, 1105–1112.
21. Lou, R.; Wu, S.; Lv, G.; Yang, Q. Energy and resource utilization of deinking sludge pyrolysis. Appl. Energy 2012, 90, 46–50.
22. Bergman, P.C.A.; Boersma, A.R.; Zwart, R.W.R.; Kiel, J.H.A. Torrefaction for Biomass Co-Firing in Existing Coal-Fired Power Stations; ECNC050132005; Energy Research Centre of the Netherlands (ECN), ECN Biomass: Petten, The Netherlands, 2005.
23. Fonts I.; Juan A.; Gea, G.; Murillo, M.B.; Sanchez, J.L. Sewage sludge pyrolysis in fluidized bed, 1: Influence of operational conditions on the product distribution. Ind. Eng. Chem. Res. 2008, 47, 5376–5385.
24. Fonts, I.; Kuoppala, E.; Oasmaa, A. Physicochemical properties of product liquid from pyrolysis of sewage sludge. Energy Fuel 2009, 23, 4121–4128.
25. Jolliet, O.; Margni, M.; Charles, R.; Humbert, S.; Payet, J.; Rebitzer, G.; Rosenbaum, R. IMPACT 2002+: A new life cycle impact assessment methodology. Int. J. Life Cycle Assess. 2003, 8, 324–330.
26. Tyagi, V.K.; Lo, S.L. Enhancement in mesophilic aerobic digestion of waste activated sludge by chemically assisted thermal pretreatment method. Bioresour. Technol. 2012, 119, 105–113.
27. Environmental Protection Administration of Taiwan Home Page. Available online: http://ivy4.epa.gov.tw/swims/swims_net/index.aspx (accessed on 7 January 2012).
28. Houillon, G.; Jolliet, O. Life cycle assessment of processes for the treatment of wastewater urban sludge: energy and global warming analysis. J. Clean Prod. 2005, 13, 287–299.

29. Lundin, M.; Olofsson, M.; Pettersson, G.J.; Zetterlund, H. Environmental and economic assessment of sewage sludge handling options. Resour. Conserv. Recycl. 2004, 41, 255–278.
30. Svanstrom, M.; Froling, M.; Olofsson, M.; Lundin, M. Environmental assessment of supercritical water oxidation and other sewage sludge handling options. Waste Manag. Res. 2005, 23, 356–366.

There are several supplemental files that are not available in this version of the article. To view this additional information, please use the citation on the first page of this chapter.

PART VI

GASIFICATION

CHAPTER 11

An Experimental and Numerical Investigation of Fluidized Bed Gasification of Solid Waste

SHARMINA BEGUM, MOHAMMAD G. RASUL,
DELWAR AKBAR, AND DAVID CORK

11.1 INTRODUCTION

Solar, wind, tides, geothermal and hydroelectric are very popular renewable energy sources. However, there is another significant source of energy that is created by our everyday activities across the world known as biomass. Biomass plays a vital role in the production of fuel or electricity. The potential of renewable energy for urban and rural development, liquid fuel replacement and greenhouse gas reduction are current concerns all over the world. Stucley et al. [1] examined the use of biomass to generate electricity and produce liquid transport fuels in Australia. The main focus of their study was on biomass from forestry, particularly new forestry that may also provide other environmental benefits in Australia's dry land regions.

An Experimental and Numerical Investigation of Fluidized Bed Gasification of Solid Waste. © Begum S, Rasul MG, Akbar D, and Cork D. Energies 7,1 (2014), doi:10.3390/en7010043. Licensed under a Creative Commons Attribution 3.0 Unported License, http://creativecommons.org/licenses/by/3.0/.

According to renewable energy perception, biomass can be defined as [1]: "Recent organic matter originally derived from plants as a result of the photosynthetic conversion process, or from animals, and which is destined to be utilized as a store of chemical energy to provide heat, electricity, or transport fuels."

Biomass resources include wood from plantation forests, residues from agricultural or forest production, and organic waste by-products from industry, domesticated animals, and human activities [1,2]. According to biomass definition, solid waste (SW) is a proven natural resource for renewable energy. Energy can be recovered from SW through various technologies, such as:

- Combustion, which is a rapid chemical reaction of two or more substances, is commonly called burning. In practical combustion systems the chemical reactions of the major chemical species, carbon (C) and hydrogen (H_2) in the fuel and oxygen (O_2) in the air, are fast at the prevailing high temperatures (approximately, greater than 900 °C) because the reaction rates increases exponentially with temperature.
- Pyrolysis and gasification, where the fuel is heated with little or no O_2 to produce "syngas" which can be used to generate energy or as a feedstock for producing methane (CH_4), chemicals, biofuels or H_2.

Amongst a number of conversion routes of waste to energy technology, gasification plays a vital role. Gasification uses partial oxidation, in contrast to combustion which uses excess air. It produces a combustible gas which is a mixture of carbon monoxide (CO), H_2 and CH_4. Biomass gasification plants are in the early commercial stage of development. When used to produce electricity, there are considerable gains associated with scale of plant. At 1 MW electrical (MWe) an updraft gasifier has an efficiency of 10% to 20%, while a 10 MWe fluid bed gasifier has an efficiency of 25% to 35%, and a 100 MWe entrained flow or pressurized circulating fluid bed gasifier has an efficiency of 40% to 50% [2].

A number of studies have been performed on the gasification process. Doherty et al. [3] developed a simulation model of a circulating fluidized bed (CFB) gasifier using Advanced System for Process ENgineering (Aspen) Plus. They calibrated the model against experimental data and investigated the effects of varying equivalence ratio (ER), temperature, level of air preheating, biomass moisture and steam injection on product gas composition, gas heating value, and cold gas efficiency. Nikoo and Mahinpey [4] developed a fluidised bed gasification model for biomass addressing both hydrodynamic parameters and reaction kinetic in their model. Kumar et al. [5] studied on simulation of corn stover and distillers grains gasification where they developed to simulate the performance of a lab-scale gasifier and predict the flowrate and composition of product from given biomass composition and gasifier operating conditions using Aspen Plus software. They applied mass balance, energy balance, and minimization of Gibbs free energy during the gasification to determine the product gas composition. Abdelouahed et al. [6] presented a detailed modeling of biomass gasification in dual fluidized bed (DFB) reactors under Aspen plus. In their model, the DFB was divided into three modules according to the main chemical phenomena: biomass pyrolysis, secondary reactions and char combustion. Most recently, Mavukwana et al. [7] developed a simulation model of sugarcane bagasse gasification, and Francois et al. [8] reported process modeling of a wood gasification combined heat and power plant using Aspen Plus.

This paper presents a detailed investigation of SW gasification, both experimentally and numerically. Experimental investigations have been done in a pilot-scale gasification plant. A numerical simulation model has been developed using the Aspen Plus software package. The model has been validated with the experimental data. The model will be very useful for the professionals, researchers and industry people involved in waste to energy technologies.

11.2 EXPERIMENTAL INVESTIGATIONS

The experimental study was performed using a pilot-scale gasification plant which is used for energy recovery from SW. The layout of the gasification plant is shown in Figure 1. The gasification plant is composed of four modules: a waste pre-processing unit, the gasification/oxidation chambers, the energy recovery section and finally the flue gas cleaning section. In the pre-processing module the waste is sorted, grinded, shredded, stored and dried with the purpose of obtaining a gasification-friendly feed material, free of metals, glass and plastic bottles. The main components of the plant are: waste receiving area, magnetic separator, shredder, hopper, reactor, hot-oil tank and bag house (Figure 1).

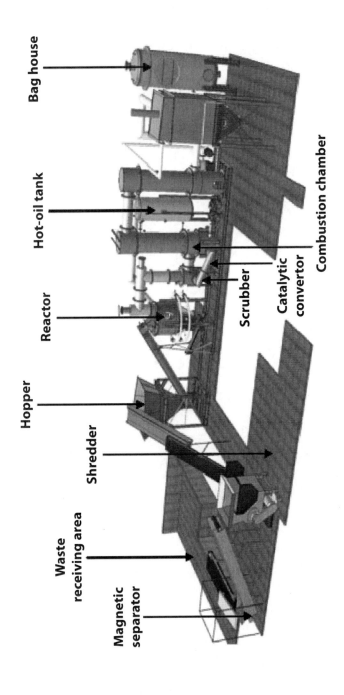

FIGURE 1: Layout of gasification plant (Reprinted with permission from The Corky Group, 2010 [9]).

Experimental and Numerical Investigation of Fluidized Bed Gasification

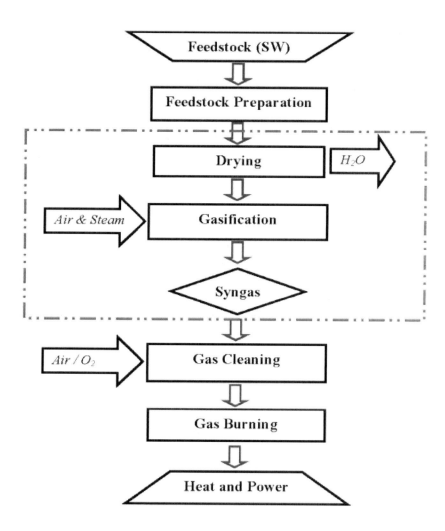

FIGURE 2: Schematic diagram of solid waste (SW).

The gasification process involves a number of steps. A schematic diagram of the gasification process is shown in Figure 2. Wood was used as SW feedstock. Feedstocks are collected from neighboring city or regional councils. Collected wastes are prepared for gasification. Feedstock preparation includes feedstock sorting, preparation and shredding. In this step, all unexpected materials have been removed. The feedstock is manually hand sorted to ensure that recycling was maximized and the shredder is protected from rocks, glass and metallic items. This ensures that no dangerous items enter the reactor vessel. All feed are shredded before being placed into the hopper and subsequent screw conveyor. As ideal sorting is complex, reactors have scrubbing systems to manage any slippage through the sorting process. The scrubbing systems are not a substitute for lack of sorting. A slippage rate of 2% is considered normal.

To increase gasification performance, prepared feedstocks need to be dried to reduce moisture through the drier. Gasification process is placed in reaction vessel in the presence of air and steam which produce raw syngas. Syngas cleaning is done to remove the pollution from produced raw syngas. Burning is done in an internal combustion engine depending on its requirement.

The reaction vessel of the plant is used to convert waste to a small amount of char and ash along with a large amount of syngas (called gasification). In the reaction vessel:

- Waste's surface moisture is dried, along with all of the inherent moisture.
- Waste is further heated.
- Waste generates volatiles (oils, tars and CH4 which appear as yellow smoke).
- Partial combustion (yellow smoke) generates heat and evaporates plastics (if present).
- Char is gasified by reacting with O_2, water or steam (H_2O) and carbon dioxide (CO_2) to form H_2 and CO.

The experimental reaction vessel had a nominal volume of 2.5 m^3, a waste and char capacity of approximately 1.75 m^3 and a freeboard capacity of 0.75 m^3. The freeboard is the space above the fluidised bed where the gas slows down after leaving the bed and most of the solids fall back into the bed. The waste is fluidised in the reactor by a gaseous mixture of waste gas [burnt syngas—nitrogen (N_2), CO_2 and H_2O] and fresh air (N_2, O_2) [9].

The wastes are heated and dried in the drier which uses hot oil at 220 °C. The SWs are heated in the range of 700–1100 °C in a reduced O_2 environment in the reactor. In the reactor, raw syngas is produced which is a mixture of CO, CO_2, H_2, H_2O, N_2 and CH_4. The raw syngas is contaminated with tar hydrogen sulphide (H_2S), ammonia (NH_3) and trace hydrochloric acid (HCl). The fine SW leaves the reactor and settles on the top of the scrubber and forms a network of fine C char. The tar gets trapped on the char network and on burnt dolomite at the top of the scrubber and is subsequently broken down into C, CO, and H_2 and H_2O. The H_2S and NH_3 are converted to sulphur dioxide (SO_2), nitric oxide (NO) and H_2O in the scrubber. The dolomite mops up the acid gases like SO_2, NO, NO_2 and HCl. O_2 is added at the scrubber to maintain the scrubber temperature at about 600 °C [9].

TABLE 1: Compounds of syngas at different stages of gasification.

Compound	Raw syngas (%)	Scrubbed syngas (%)	Dewatered syngas (%)
Hydrogen (H_2)	19.0	20.0	20.8
Carbon dioxide (CO_2)	11.4	12.0	12.5
Carbon monoxide (CO)	28.5	30.0	31.3
Methane (CH_4)	1.9	2.0	2.1
Water or steam (H_2O)	5.7	6.0	2.1
Nitrogen (N_2)	28.5	30	31.3
Phenol, cresol, tars, hydrogen sulphide (H_2S) and ammonia (NH_3)	5	-	-

The hot syngas is indirectly cooled using heat transfer oil. The heated oil is used for drying the SW. The hot syngas is further indirectly cooled using fluidising gas. The heated fluidizing gas is used in the gasifier. The warm syngas is further cooled and compressed using liquid ring compressors.

This is a final scrub for acid gases and also removes about 80% of the water out of the syngas. The clean syngas is a mixture of CO, CO_2, H_2, N_2 and H_2O. The compounds of syngas found at different stages are shown in Table 1. Energy content of SW was distributed to, approximately, 65% for syngas production, 23% for char production, 6% for hot oil and remaining was lost as heat. This implies that the energy conversion efficiency for syngas production using gasification is approximately 65%.

11.3 MODEL DEVELOPMENT

A simulation model was developed and validated with experimental results. Amongst the whole experimental study, the model was developed for the main part of the gasification process (as identified by the red rectangle in Figure 2).

11.3.1 MODEL ASSUMPTIONS

Because of the influence of hydrodynamic parameters on SW gasification in a fluidized bed, both the hydrodynamic and kinetic reactions were treated simultaneously. A number of assumptions were incorporated into the Aspen Plus fluidized bed gasifier modeling [4,10–13]:

- The gasification process is steady state and isothermal (uniform bed temperature).
- SW de-volatilisation is instantaneous in comparison to char gasification.
- Volatile products mainly consist of CO, H_2, CO_2, CH_4 and H_2O.
- Gases are uniformly distributed within the emulsion phase.
- Char comprises C and ash.
- Char gasification starts in the bed and is completed in the freeboard. According to Lee et al. [14], combustion and gasification take place in the main bed region and pyrolysis in the freeboard region.
- Based on the shrinking core model, particles are considered spherical and of uniform size and the average diameter remains constant during gasification.
- The simulation is performed using power-law kinetics.

11.3.2 REACTION KINETICS

The overall gasification process starts with pyrolysis and continues with combustion and steam gasification considering the following reactions:

Combustion reactions [4,14–16]:

$$C + \alpha O_2 \rightarrow 2(1-\alpha)CO + (2\alpha-1)CO_2 \qquad (1)$$

$$C + CO_2 = 2CO + \text{(Boudouard)} \qquad (2)$$

Steam-gasification reactions [4,17]:

$$C + H_2O = CO + H_2 \qquad (3)$$

$$CO + H_2O = CO_2 + H_2 \qquad (4)$$

$$C + 2H_2O = CO_2 + 2H_2 \qquad (5)$$

$$C + \beta H_2O \rightarrow (\beta-1)CO_2 + (2-\beta)CO + \beta H_2 \qquad (6)$$

In the reaction shown in Equation (1), α is a mechanism factor [18] that differs in the range of 0.5–1 when CO_2 and CO is carried away from the char particle during combustion of char. α is a function of temperature and average diameter of char particles. In the reaction shown in Equation (5), the term $(2 - \beta)/\beta$ represents the fraction of the steam consumed by the reaction shown in Equation (2) and $2(\beta - 1)/\beta$ represents the fraction of steam consumed by the reaction shown in Equation (4). Matsui et al. [15] experimentally determined β to be in the range of 1.1–1.5 at 750–900 °C. The values of α and β for fluidized bed model of SW gasification were considered as 0.9 and 1.4, respectively, as determined by Nikoo and Mahinpey [4].

According to Lee et al. [14], the reaction rate equations defined for the mentioned reactions are as follows:

$$\frac{dX_{CO}}{dt} = k_{CO} exp\left(\frac{-E_{CO}}{RT}\right) p_{O_2}^n (1 - X_{CO})^{2/3} \tag{7}$$

$$\frac{dX_{SG}}{dt} = k_{SG} exp\left(\frac{-E_{SG}}{RT}\right) p_{H_2O}^n (1 - X_{SG})^{2/3} \tag{8}$$

$$r_c = \left(\frac{dX_{CO}}{dt} + \frac{dX_{SG}}{dt}\right) \times \frac{\rho_c \varepsilon_s Y_c}{M_C} \tag{9}$$

Walker et al. [19] and Dutta and Wen [20] considered parameter n to be equal to 1.0 in Equations (7) and (8). For the steam-gasification reaction, Kasaoka et al. [21] and Chin et al. [22] reported different numbers for n, but it is actually 1.0 in the steam partial pressure range of 0.25–0.8 atm. Kinetic parameters can be found in Table 2 [4]. All the symbols for the reaction kinetics have been defined in the nomenclature.

TABLE 2: Kinetic parameters.

Process	E/R (K)	k (s^{-1} atm^{-1})
Combustion	13,523	0.046
Gasification with steam	19,544	6,474.7

11.3.3 HYDRODYNAMIC ASSUMPTIONS

The following assumptions were considered in simulating the hydrodynamics [4,23]:

- Fluidized bed reactor comprises with two regions: bed and freeboard.
- The fluidisation condition in the bed is maintained in the bubbling regime.
- With the increasing height, the solids volume fraction decreases, similar to the grouping of bubbles in with solid particles returning to the bed.
- With increasing height, the volumetric flow rate of gas increases corresponding to the production of gaseous products.

- Solid particles mixing, such as consisting of ash, char particles, and bed material, are considered perfect.
- With constant hydrodynamic parameters, the reactor is divided into a finite number of equal elements.
- As the fluidised bed is considered one-dimensional, any variations in conditions are considered to occur only in the axial direction.

11.3.4 BED HYDRODYNAMICS

According to Kunii and Levenspiel [24], the minimum fluidisation velocity of fine particles can be determined by:

$$u_{mg} = \frac{33.7\mu}{\rho_g d_p}\left(\sqrt{1 + 3.59 \times 10^{-5} Ar} - 1\right) \tag{10}$$

$$Ar = \frac{d_p^3 \rho_g (\rho_s - \rho_g) g}{\mu^2} \tag{11}$$

To determine the volume fraction occupied by bubbles in a fluidized bed, the following correlations developed by Babu et al. [25] are used:

$$B = 1.0 + \frac{10.978(u - u_{mf})^{0.738} \rho_s^{0.376} d_p^{1.006}}{u_{mf}^{0.937} \rho_g^{0.126}} \tag{12}$$

$$\varepsilon_b = 1 - \frac{1}{B} \tag{13}$$

Here, u is the superficial velocity which is not a constant parameter due to the gas production resulting from heterogeneous and homogeneous reactions. The importance of considering varying superficial gas velocity in obtaining results with higher precision in simulation is demonstrated by Yan et al. [26]. According to Kunii and Levenspiel [24], the bed void fraction is calculated using the following equation:

$$\varepsilon_f = \varepsilon_b + (1 - \varepsilon_b)\varepsilon_{mf} \qquad (14)$$

11.3.5 FREEBOARD HYDRODYNAMICS

The volume fraction of solid at various levels z in the freeboard falls off exponentially from the value at the bed surface as calculated by the following equation [27]:

$$1 - \varepsilon_{fb} = \left(1 - \varepsilon_f\right)\exp(-az) \qquad (15)$$

Here, a is a constant. Kunii and Levenspiel [24] present a graph that correlates the constant a with superficial gas velocity and particle size. The graph is useful in the following range: $u \leq 1.25$ m/s:

$$a = 1.8/u \qquad (16)$$

11.3.6 PROCESS SIMULATOR: ASPEN PLUS

A number of processes modeling software package are used to develop computational model of fluidized bed gasification process and to perform simulation and validation of the model. Generally, researchers and professionals uses, Aspen Plus, Computational Fluid Dynamics (CFD, composed of GAMBIT and FLUENT), ChemCAD and MatLab software packages to develop and optimize the gasification model. Mhilu [28] conducted a study on modeling performance of high-temperature biomass gasification process using MatLab. In this study, the derived model equations were computed using the MAPLE process simulation code in MatLab. Sofialidis and Faltsi [29] studied on simulation of biomass gasificationin fludixed beds using CFD approach. Although CFD is a powerful software, its programs have high computational requirements. On the other hand, Aspen Plus, a familiar, proven and acceptable processes modeling software, used in the fields of SW and coal gasification, oil industry and others. It is a powerful process modeling tool which offers to include customize user models to embed with built in Aspen reactor blocks. Since it contains a large property database for conventional compounds and convergence algorithms for solving minimization problems, Aspen Plus is selected to develop a gasification model. Many researchers have used Aspen Plus to develop gasification and downstream models for SW, coal and other biomass [3–6]. The SW gasification model can be separated in two ways according to Lu et al. [30]:

- Kinetic model: is capable of simulating the reaction conditions at different times and sites which is suitable for reactor amplification design and operation parameters optimisation.
- Equilibrium model: predicts only end reaction product distribution but no idea is provided about instantaneous product distribution along with geometric dimensions.

In this study, both reaction kinetics parameters and bed hydrodynamics aspects are considered to develop the model. The development of a fluidized bed gasification model through Aspen Plus involves the following steps:

- steam class specification;
- property method selection;
- system component specification (from databank) and identifying conventional and non-conventional components;
- defining the process flowsheet (using unit operation blocks and connecting material and energy streams);
- specifying feed streams (flow rate, composition and thermodynamic condition), and
- specifying unit operation blocks (thermodynamic condition, chemical reactions, etc.).

Aspen Plus lacks a built-in library of models with which to develop a customized fluidized bed gasification model. However, Aspen Plus provides the facility for the user to input their own models using FORTRAN/ Excel codes and reactions nested within the input file.

11.3.7 MODEL DESCRIPTION

The fluidized bed gasification model is comprised of a number of Aspen reactor blocks. In order to show the overall gasification process, there are different phases considered in the Aspen Plus simulation, these being drying, pyrolysis (decomposition), volatile reactions, char gasification, and gas-solid separation. The process flowchart and an Aspen Plus simulation flowsheet of biomass gasification are shown in Figures 3 and 4, respectively.

Feed (SW), dry-feed and ash are specified as a non-conventional component in Aspen Plus and defined in the simulation model by using the ultimate and proximate analysis. The characteristics of feed are shown in Table 3.

TABLE 3: Characteristics of SW (wood).

Moisture content (MC) (%)		25
Proximate analysis (mass %) (Dry basis)	Volatile matter (VM)	82.6
	Fixed carbon (FC)	16.3
	Ash	1.1
Ultimate Analysis (mass %)	Carbon (C)	49.8
	H	26.1
	Oxygen (O_2)	33.9
	N	20.2
	Sulphur (S)	0.1
Average particle size (mm)		0.25–0.95
Char density (kg/m^3)		1400
Gross specific energy (MJ/kg)		18.6

The input parameters of the corresponding gasifier operating conditions were similar to experimental measurements, are given in Table 4.

Methods and processes used for developing the Aspen Plus model is briefly described below.

TABLE 4: Gasifier operating parameters.

Feed	Flow rate (kg/h)	4.5
	Pressure (MPa)	0.3
	Temperature (°C)	25
Air	Flow rate (kg/h)	4.5
	Pressure (MPa)	0.3
	Temperature (°C)	350
	Air/fuel ratio	1
Steam	Flow rate (kg/h)	27
	Pressure (MPa)	0.3
	Temperature (°C)	200
	Steam/fuel ratio	6
Reaction vessel volume	Total (m^3)	2.5
	Waste/char capacity (m^3)	1.75
	Freeboard capacity (m3)	0.75
Gasifier	Pressure (MPa)	0.3
	Temperature (°C)	700–1100
Dryer	Pressure (MPa)	0.3
	Temperature (°C)	400
Decomposition	Pressure (MPa)	0.3
	Temperature (°C)	400

11.3.7.1 PHYSICAL PROPERTY METHOD

The IDEAL property method was set for this simulation in which ideal behaviors are assumed, such as systems at vacuum pressures and isomeric systems at low pressures. In the vapor phase, small deviations from the ideal gas law are allowed. These deviations occur at low pressures (either below atmospheric pressure, or at pressures not exceeding 2 bar) and very high temperatures. Ideal behavior in the liquid phase is exhibited by molecules with either very small interactions or interactions that cancel each other out. The IDEAL property method is generally used for systems with and without non-condensable components. In this method, permanent gases can be dissolved in the liquid [31].

The steam class was set as MIXED, NC and PSD (MIXNCPSD). MIXNCPSD represents that, nonconventional solids are present, with a particle size distribution. The NC properties: Enthalpy and Density model was selected as HCOALGEN and DCOALIGT, respectively, for both feed, dry-feed and ash which are non-conventional components. HCOALGEN is the general coal/SW model for computing enthalpy in the Aspen Physical Property System which includes a number of different correlations for: heat of combustion, heat of formation and heat capacity. The density model, DCOALIGT, gives the true (skeletal or solid-phase) density of coal/SW on a dry basis using ultimate and sulfur analyses.

11.3.7.2 MODEL SEQUENCE

A number of Aspen Plus blocks were used to complete the overall gasification process. The main processes were simulated by three reactors in Aspen Plus: RYield, RGibbs and RCSTR. Additionally, a MIXER and a number of SEPERATOR blocks were incorporated in the simulation model to complete the entire process. The whole gasification consists of four processes, namely drying, decomposition, volatile reaction and char gasification and combustion.

11.3.7.3 DRYING

Reducing moisture from feed plays important role to improve gasifier performance. An Aspen Plus block, RYield (block ID: DRYER in Figure 4) was used to simulate the drying process of SW. The SW is fed into the block, and the water bound in SW is vaporized in this block. The yield of gaseous water is determined by the water content in the proximate analysis of SW. The moisture content of the SW is 25%, therefore, the mass yield of gaseous water is set as 25%, based on the assumption that the physically bound water is vaporized completely in this process. The mass yield of dried SW is correspondingly equal to 100% − 25% = 75%. After the drying process, the gaseous water and dried SW flow into a gas and solid separator, SEP0. The separated gaseous water is drained out from the process and the separated dried SW goes on to the next block for the decomposition of dried feed.

11.3.7.4 DECOMPOSITION

RYield (block ID: DCOMP in Figure 4), an Aspen Plus yield reactor, was used to decompose the feed in the simulation. RYield is used when reaction stoichiometry and reaction kinetics are unknown or unimportant but the component yield distribution is known. In this step, SW is converted into its constituent components H_2, O_2, C, sulphur (S), N_2 and ash by specifying the yield distribution according to the feed's ultimate analysis.

11.3.7.5 VOLATILE REACTIONS

An Aspen Plus reactor, RGibbs (block ID: GASIFY in Figure 4), uses Gibbs free energy minimization with phase splitting to calculate equilibrium. This reactor does not require specifying the reaction stoichiometry, but reactor temperature and pressure is known from experiment. RGibbs is capable of calculating the chemical equilibrium between any number of conventional solid components and the fluid phases [32].

In this study, RGibbs was used for volatile combustion. SW mainly consists of C, O_2, H_2, N_2, chlorine (Cl), S, moisture and ash. Here, C will partly compose the gas phase to take part in de-volatilization and the remaining part of C comprises the solid phase (char) and consequently results in char gasification. A separator was used before RGibbs reactor to separate the volatile materials and solids from the decomposed components. RGibbs reactor performs the volatile reactions of separated volatile materials.

230 Solid Waste as a Renewable Resource: Methodologies

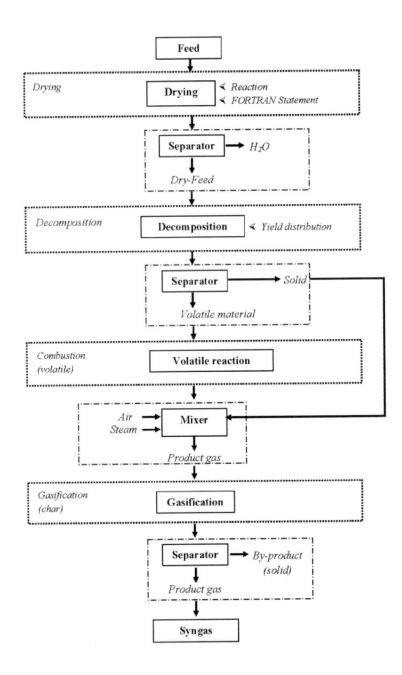

FIGURE 3: Process flowchart of fluidized bed gasification.

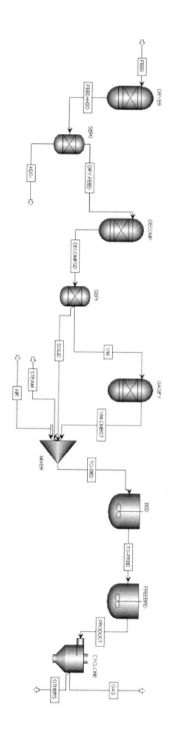

FIGURE 4: Advanced System for Process ENgineering (Aspen) Plus simulation flowsheet.

11.3.7.6 CHAR GASIFICATION

RCSTR, one of the vital reactors in Aspen Plus, rigorously models continuously stirred tank reactors. This reactor is capable of modelling one, two or three phase reactors. RCSTR assumes perfect mixing in the reactor, that is, the reactor contents have the same properties and composition as the outlet stream. This reactor handles kinetic and equilibrium reactions as well as reactions involving solids. The user can provide the reaction kinetics through the built-in Reactions models or through a user-defined FORTRAN/Excel subroutine [32].

In this study, the reactor, RCSTR (block ID: BED and FREEBRD in Figure 4) was used to perform char gasification using reaction kinetics as discussed previously. The reaction kinetics were integrated by written FORTRAN and Excel code via the CALCULATOR block. The reactor was divided into bed and freeboard regions using hydrodynamic parameters where each region is simulated by one RCSTR reactor. Using FORTRAN code, each RCSTR is divided into a series of CSTR reactors with equal volume. Bed and Freeboard reactors require matching a number of variables, some specified at the top and others at the bottom of the gasifier. This feature causes the solution process to be usually complicated and time-consuming. Therefore, from the viewpoint of directly using the built-in algorithm in Aspen Plus and then simplifying the problem, a number of RCSTR reactors in series are used to model the char gasification and combustion processes. The RCSTR reactor has the characteristic that all phases have the same temperature, which means the temperatures of solid and gas phases in the char gasification and combustion processes are equal in the model. The kinetic and hydrodynamic parameters such as fractional pressure, superficial velocity and voidage of O_2 and steam remain constant in these reactors.

11.3.7.7 SOLID SEPARATION

A splitter block, CYCLONE, was used in this model to separate C solid from the gas mixture provided by the split fraction of MIXED and NCPSD.

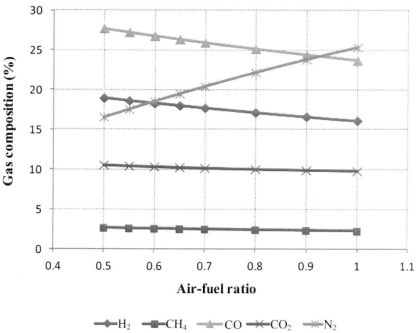

FIGURE 5: Effect of air-fuel ratio on syngas composition.

11.4 RESULTS AND DISCUSSIONS

11.4.1 MODEL VALIDATION

This section presents a comparison of simulation results with experimentally measured data using a pilot scale gasifier shown in Figure 1. The operating conditions of the gasifier model were similar to the experimental measurement as shown in Table 4. The composition of raw syngas, such as H_2, CO, CO_2, CH_4, N_2 and H_2O, was measured experimentally and is compared in Table 5 with the simulation results. It can be clearly seen from Table 5 that there is only about 3% variation between measured and simulated results. Therefore, it is fair to say that the developed model can be used for further analysis with acceptable accuracy. The validated model was used to study the effect of air-fuel ratio and steam-fuel ratio on syngas composition as discussed below.

TABLE 5: Comparison of experimental and simulation result of SW gasification.

Measurement	H_2	CO	CO_2	CH_4	N_2	H_2O	Others
Experimental (%)	19	28.5	11.4	1.9	28.5	5.7	5.0
Model (%)	17.37	25.49	10.55	2.44	27.29	9.2	7.65
Difference (%)	−1.63	−3.01	−0.85	0.54	−1.21	3.5	2.65

11.4.2 SIMULATION MODEL ANALYSIS

11.4.2.1 EFFECT OF AIR-FUEL RATIO

The effect of air-fuel ratio on product gas composition was examined. Simulation results for syngas composition (in percent) versus air-fuel ratios in the range of 0.5 to 1.0 are shown in Figure 5. The production of both H_2 and CO decreases with the increasing amount of air, while the volume of the inert gas N_2 in the syngas increases. The composition of CO_2 decreases with very small deviation and CH_4 remains almost the same with increasing air-fuel ratio. Due to the high concentration of N_2 in air, N_2 concentration increases in syngas composition with increasing air supply.

Air-fuel ratio not only represents the O_2 quantity introduced into the reactor, but also affects the gasification temperature under the condition of auto thermal operation. Higher air-fuel ratios can cause syngas quality to degrade because of an increased oxidation reaction. Alternatively, higher air-fuel ratios mean a higher gasification temperature which can accelerate the gasification and improve the product quality to a certain extent. The oxidation reaction for CO production is:

$$C + 0.5O_2 = CO \tag{17}$$

The oxidation reaction for CO_2 production is:

$$C + O_2 = CO_2 \tag{18}$$

Based on the oxidation reactions, Equations (17) and (18), CO production consumes more C for the same amount of O_2.

11.4.2.2 EFFECT OF STEAM-FUEL RATIO

The effect of steam-fuel ratio in the range of 4–10 on syngas composition is shown in Figure 6. The concentration of CO and H_2 exhibits a trend that slowly increases when the steam to fuel ratio increases. This can be explained by more steam reforming reactions of CO and H_2 taking place because of increased steam quantity. The change in CH_4 and CO_2 concentration is very small with increasing steam-fuel ratio. The C-steam reaction is highly temperature sensitive, as an increase in low steam flows cause the efficiency to increase slightly in the experiment, but it then drops for additional rises in the steam flow amount. The concentration of N_2 composition shows a significant decrease (from 20% to 13%) with increasing steam-fuel ratio as there is only a fixed amount of air supplied with increasing steam-fuel ratio.

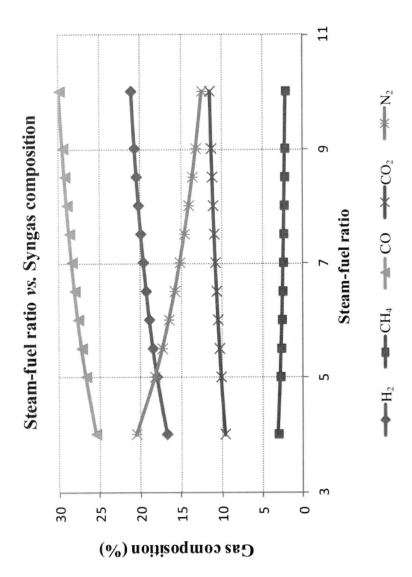

FIGURE 6: Effect of steam-fuel ratio on syngas composition.

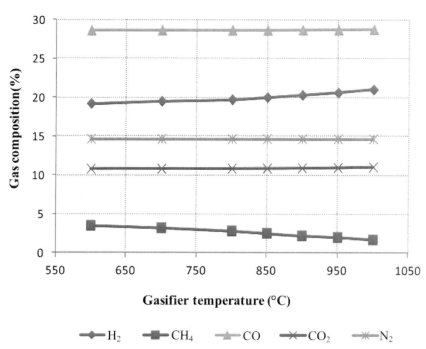

FIGURE 7: Effect of gasifier temperature on syngas composition.

11.4.2.3 EFFECT OF GASIFIER TEMPERATURE

The effect of gasifier temperature on produced syngas composition is shown in Figure 7. The temperature considered varies from 600 °C to 1000 °C. The concentration of syngas components (i.e., CO, H_2, CO_2, CH_4 and N_2) varying with a small range with increasing gasifier temperature. There are about 28%, 20%, 15%, 11% and 2% of CO, H_2, N_2, CO_2 and CH_4 produced, respectively. The overall gasification process, along with C conversion, is improved by increases in gasification temperature. The rate of increase in C conversion efficiency becomes slow at higher gasification temperatures.

In lower gasifier temperature, SW produces more unburned hydrocarbon and tar, which decreases H_2 production. In Figure 7, the higher amount of H_2 favors the backward reaction and causes prediction of lower CO_2 production in simulation related to the reaction in Equation (6). The trend of syngas composition in simulation is following the same trend as literature [4].

11.5 CONCLUSIONS

Experimental and numerical investigations of atmospheric fluidized bed gasification have been done for SW. The experimental investigation was performed using a pilot-scale gasifier plant. An Aspen Plus simulation model was developed based on the experimental setup and findings. To develop the simulation model, several Aspen Plus reactor blocks were used with a number of separators and a mixer block. Reaction kinetics and hydrodynamic equations were incorporated through FORTRAN and Excel code. The simulation model was validated with experimental results of a pilot scale SW gasification plant. A very good agreement was found between simulation and experimental results, with a maximum variation of 3%. The effects of air-fuel and steam-fuel ratio on syngas composition were simulated using the validated model. The model will assist researchers, professionals and industry people to identify the optimized conditions for SW gasification.

The experimental investigation was supported by The Corky's Group: Corkys Carbon and Combustion.

REFERENCES

1. Stucley, C.; Schuck, S.; Sims, R.; Larsen, P.; Turvey, N.; Marino, B. Biomass Energy Production in Australia, revised ed.; Rural Industries Research and Development Corporation: Canberra, Australia, 2004.
2. Dopita, M.; Williamson, R. Australia's Renewable Energy Future; Australian Academy of Science: Canberra, Australia, 2010.
3. Doherty, W.; Reynolds, A.; Kennedy, D. The effect of air preheating in a biomass CFB gasifier using ASPEN Plus simulation. Biomass Bioenergy 2009, 33, 1158–1167.
4. Nikoo, M.B.; Mahinpey, N. Simulation of biomass gasification in fluidized bed reactor using ASPEN PLUS. Biomass Bioenergy 2008, 32, 1245–1254.
5. Kumar, A.; Noureddini, H.; Demirel, Y.; Jones, D.D.; Hanna, M.A. Simulation of corn stover and distillers grains gasification with Aspen Plus. Trans. ASABE 2009, 52, 1989–1995.
6. Abdelouahed, L.; Authier, O.; Mauviel, G.; Corriou, J.P.; Verdier, G.; Dufour, A. Detailed modeling of biomass gasification in dual fluidized bed reactors under Aspen Plus. Energy Fuels 2012, 26, 3840–3855.
7. Mavukwana, A.; Jalama, K.; Ntuli, F.; Harding, K. Simulation of Sugarcane Bagasse Gasification using Aspen Plus. Presented at the International Conference on Chemical and Environmental Engineering (ICCEE), Johannesburg, South Africa, 15–16 April 2013.
8. Francois, J.; Abdelouahed, L.; Mauviel, G.; Patisson, F.; Mirgaux, O.; Rogaume, C.; Rogaume, Y.; Feidt, M.; Dufour, A. Detailed process modeling of a wood gasification combined heat and power plant. Biomass Bioenergy 2013, 51, 68–82.
9. Gasifier Operator Manual; The Corky's Group: Mayfield, Australia, 2010.
10. Sadaka, S.S.; Ghaly, A.E.; Sabbah, M.A. Two phase biomass air-steam gasification model for fluidized bed reactors: Part I—Model development. Biomass Bioenergy 2002, 22, 439–462.
11. Lv, P.M.; Xiong, Z.H.; Chang, J.; Wu, C.Z.; Chen, Y.; Zhu, J.X. An experimental study on biomass air–steam gasification in a fluidized bed. Bioresour. Technol 2004, 95, 95–101.
12. Buekens, A.G.; Schoeters, J.G. Modelling of Biomass Gasification. In Fundamentals of Thermochemical Biomass Conversion; Overend, R.P., Milne, T.A., Mudge, K.L., Eds.; Elsevier Applied Science Publishers: London, UK, 1985; pp. 619–689.
13. Ergudenler, A. Gasification of Wheat Straw in a Dual-Distributor Type Fluidized Bed Reactor. In Ph.D. Thesis; Technical University of Nova: Scotia, Halifax, NS, Canada, 1993.
14. Lee, J.M.; Kim, Y.J.; Lee, W.J.; Kim, S.D. Coal-gasification kinetics derived from pyrolysis in a fluidized-bed reactor. Energy 1998, 23, 475–488.

15. Liu, G.-S.; Niksa, S. Coal conversion submodels for design applications at elevated pressures. Part II. Char gasification. Prog. Energy Combust. Sci 2004, 30, 679–717.
16. Montagnaro, F.; Salatino, P. Analysis of char-slag interaction and near-wall particle segregation in entrained-flow gasification of coal. Combust. Flame 2010, 157, 874–883.
17. Matsui, I.; Kunii, D.; Furusawa, T. Study of fluidized bed steam gasification of char by thermogravimetrically obtained kinetics. J. Chem. Eng. Jpn 1985, 18, 105–113.
18. Rajan, R.R.; Wen, C.Y. A comprehensive model for fluidized bed coal combustors. AIChE J 1980, 26, 642–655.
19. Walker, P.L.J.; Rusinko, F.J.; Austin, L.G. Gas reactions of carbon. Adv. Catal 1959, 11, 133–221.
20. Dutta, S.; Wen, C.Y. Reactivity of coal and char 2. In oxygen-nitrogen atmosphere. Ind. Eng. Chem. Process Des. Dev 1977, 16, 31–36.
21. Kasaoka, S.; Skata, Y.; Tong, C. Kinetic evaluation of the reactivity of various coal chars for gasification with carbon dioxide in comparison with steam. Ind. Eng. Chem 1985, 25, 160–175.
22. Chin, G.; Kimura, S.; Tone, S.; Otake, T. Gasification of coal char with steam. Part 2. Pore structure and reactivity. Ind. Eng. Chem 1983, 23, 113–120.
23. Athar, M. Simulation of Coal Gasification in Circulating Fludized Bed (CFB) Reactor. Master's Thesis, University of Engineering & Technology, Lahore, Pakistan, 2009.
24. Kunii, D.; Levenspiel, O. Fluidization Engineering, 2nd ed.; Butterworth-Heinemann: Newton, MA, USA, 1991.
25. Babu, S.P.; Shah, B.; Talwalker, A. Fluidization correlations for coal gasification materials-minimum fluidization velocity and bed expansion ratio. AIChE Symp. Ser 1978, 74, 176–186.
26. Yan, H.M.; Heidenreichayb, C.; Zhanga, D.K. Mathematical modelling of a bubbling fluidized-bed coal gasifier and the significance of "net flow". Fuel 1998, 77, 1067–1079.
27. Lewis, W.K.; Gilliland, E.R.; Lang, P.M. Entrainment from fluidized beds. Chem. Eng. Prog. Symp. Ser 1962, 58, 65–78.
28. Mhilu, C.F. Modeling performance of high-temperature biomass gasification process. ISRN Chem. Eng. 2012, 2012. [CrossRef]
29. Sofialidis, D.; Faltsi, O. Simulation of biomass gasification in fluidized beds using computational fluid dynamics approach. Therm. Sci 2001, 5, 95–105.
30. Lü, P.; Kong, X.; Wu, C.; Yuan, Z.; Ma, L.; Chang, J. Modeling and simulation of biomass air-steam gasification in a fluidized bed. Front. Chem. Eng. China 2008, 2, 209–213.
31. Aspen Physical Property System V7.2. Available online: http://www.aspentech.com (accessed 2 August 2013).
32. Aspen Plus. Aspen Plus User Models V7.3. Available online: http://support.aspentech.com/Public/Documents/Engineering/Aspen%20Plus/V7.3/AspenPlusUserModelsV7_3-Ref.pdf (accessed 2 August 2013).

CHAPTER 12

Gasification of Plastic Waste as Waste-to-Energy or Waste-to-Syngas Recovery Route

ANKE BREMS, RAF DEWIL, JAN BAEYENS, AND RUI ZHANG

12.1 INTRODUCTION: PSW AND ITS TREATMENT OPTIONS

Plastics are light-weight, durable, and versatile, allowing their incorporation into a diverse range of applications. In recent years, the environmental, social and economic impact of plastics has been the topic of the political agenda, with a focus on sustainable production and the decoupling of adverse environmental effects from waste generation. The disposal of waste plastics has become a major worldwide environmental problem. The USA, Europe and Japan generate about 55 million tons of post consumer plastic waste [1]. These waste products were previously dumped in landfill sites, a non-sustainable and environmentally questionable option. The number of landfill sites and their capacity are moreover decreasing rapidly and in most countries the legislation on landfills is becoming increasingly stringent.

Gasification of Plastic Waste as Waste-to-Energy or Waste-to-Syngas Recovery Route. © Brems A, Dewil R, Baeyens J, Zhang R. Natural Science 5,6 (2013), DOI:10.4236/ns.2013.56086. Licensed under a Creative Commons Attribution License, http://creativecommons.org/licenses/by/3.0/.

New sustainable processes have emerged, i.e. 1) the advanced mechanical recycling of post-consumer plastic waste as virgin or second grade plastic feedstock; and 2) thermal treatments to recycle the waste as virgin monomer, as synthetic fuel gas, as hydrocarbon feedstock, or as a heat source (incineration with energy recovery). These processes avoid land filling, where the non-biodegradable plastics remain a lasting environmental burden. Plastic solid waste (PSW) treatment can be divided in four methods, as illustrated in Figure 1, each individual method providing a unique set of advantages making it particularly suited and beneficial to a specific location, application or product requirement [2,3]. The purpose of recycling is to minimize the consumption of finite natural resources, and this is especially relevant in the case of plastics which account for the use of 4%–8% of the global oil production [3]: re-using plastics is the required course of action, with the additional benefit of reducing emissions associated with plastic production [3]. The most appropriate recovery method is chosen considering the environmental, economic and social impact of a particular technique. Figure 1 illustrates the position of each recycling method within the production chain.

12.2 THERMO-CHEMICAL RECYCLING OF PSW

Thermo-chemical recycling refers to advanced technology processes which convert plastic materials into smaller molecules, usually liquids or gases, which are suitable for use as a feedstock for the production of new petrochemicals and plastics [1]. Products have moreover been proven to be useful as fuel. The technology behind its success is the de-polymerisation process that can result in a very profitable and sustainable industrial scheme, providing a high product yield and a minimum residual waste. Processes of pyrolysis, gasification, hydrogen-nation, and steam/catalytic cracking have been previously reported upon [1].

Recently, much attention has been paid to chemical recycling (mainly pyrolysis, gasification and catalytic degradation) as a method of producing various hydrocarbon fractions from PSW. By their nature, a number of polymers are advantageous for such treatment. Thermolysis is the treatment of PSW in the presence of heat at controlled temperatures and under a controlled environment. Thermolysis processes can be divided into pyrolysis (thermal cracking in an inert atmosphere), gasification (in the substoichiometric presence of air usually leading to CO and CO_2 production) and hydrogenation (hydrocracking) [4].

Pyrolysis can be successfully applied to Polyethylene teraphthalate (PET), polystyrene (PS), polymethylmetacrylate (PMMA) and certain polyamides such as nylon, efficiently depolymerising them into constitutive monomers [5-7]. Polyolefins, and in particular polyethylene (PE), has been targeted as a potential pyrolysis feedstock for fuel (gasoline) production, or to produce waxes as feedstock for synthetic lubricants, albeit with a limited success.

The development of value added recycling technologies is highly desirable as it would increase the economic incentive to recycle polymers. Several methods for thermochemical recycling are presently in use, such as direct gasification, and degradation by liquefaction [8]. Various degradation methods for obtaining petrochemicals are presently under investigation, and conditions suitable for pyrolysis and gasification are being researched extensively [9].

Catalytic cracking and reforming facilitate the selective degradation of waste plastics. The use of solid catalysts such as silica-alumina, ZSM-5 or zeolites, effectively converts polyolefins into liquid fuel, giving lighter fractions as compared to thermal cracking.

Gasification has recently been attracting increased attention as thermo-chemical recycling technique. Its main advantage is the possibility of treating heterogeneous and contaminated polymers with limited use of pre-treatment, whilst the production of syngas creates different applications in synthesis reactions or energy utilisation. Gasification has been widely studied and applied for biomass and coal, with results reported and published in literature. The application for the treatment of plastic solid waste is less documented, although the number of publications increases exponentially.

Figure 2 shows different thermolysis schemes, main technologies and their main products obtained, as initially presented by Mastellone [10].

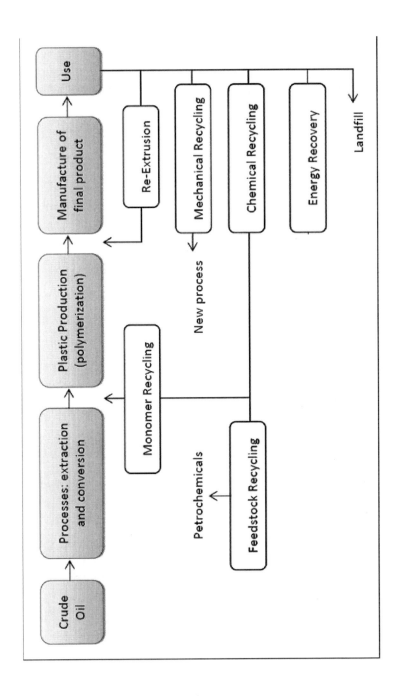

FIGURE 1: Schematic of recycling methods and their position within the processing line (Adapted from [2]).

Gasification of Plastic Waste as Waste-to-Energy 245

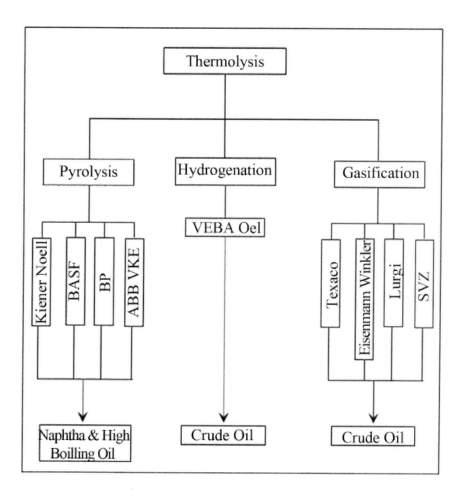

FIGURE 2: Different thermolysis schemes with reference to the main technologies [10].

12.3 GASIFICATION

12.3.1 GASIFICATION OF PSW

Gasification or partial oxidation of plastic waste is commonly operated at high temperatures (>600°C - 800°C). Air (or oxygen in some applications) is used as a gasification agent, and the air factor is generally 20% - 40% of the amount of air needed for the combustion of the PSW. The process essentially oxidizes the hydrocarbon feedstock in a controlled fashion to generate the endothermic depolymerisation heat. The primary product is a gaseous mixture of carbon monoxide and hydrogen, with minor percentages of gaseous hydrocarbons also formed. This gas mixture is known as syngas and can be used as a substitute for natural gas or in the chemical industry as feedstock for the production of numerous chemicals. For most of the PSW components, the ash and char production is limited [3,11-13].

A hydrogen production efficiency of 60%–70% from polymer waste has been reported [14]. Gasification is an attractive alternative to direct incineration of PSW, since it reduces the formation of dioxins and aromatic compounds.

Gasification efficiently utilizes the chemical energy and recoverable raw materials inherent in unsorted domestic waste, industrial and special waste (e.g. medical waste), and is capable of transforming almost all of the total waste input into technically usable raw materials and energy [12].

Co-gasification of biomass with polymers has also been shown to increase the amount of hydrogen produced while the CO content is reduced [15].

A 40 MW fluidized bed gasifier has been installed by Corenso in Varkaus (Finland) for processing polyethylene plastics with metallic aluminium recovery from recycling of Tetrapak cartons.

A typical circulating fluidized bed gasifier is illustrated in Figure 3, with the syngas being thermally used in a boiler arrangement [16].

Air is mostly used as a gasification agent: the main advantage of using air instead of O_2 alone is to simplify the process and reduce the cost. But a disadvantage is N_2 being present causing the reduction in the calorific value of resulting syngas due to the dilution.

Gasification of Plastic Waste as Waste-to-Energy 247

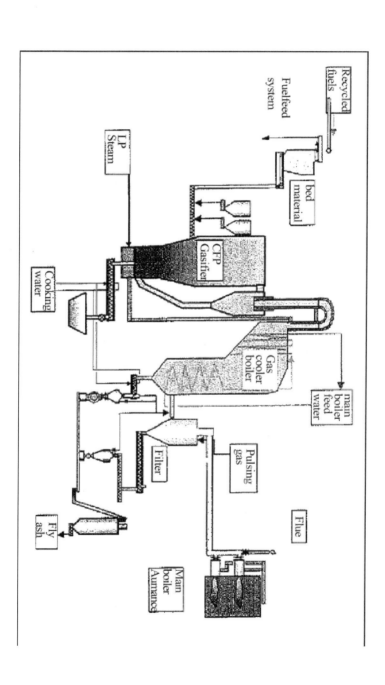

FIGURE 3: Fluidized bed gasification plant [16].

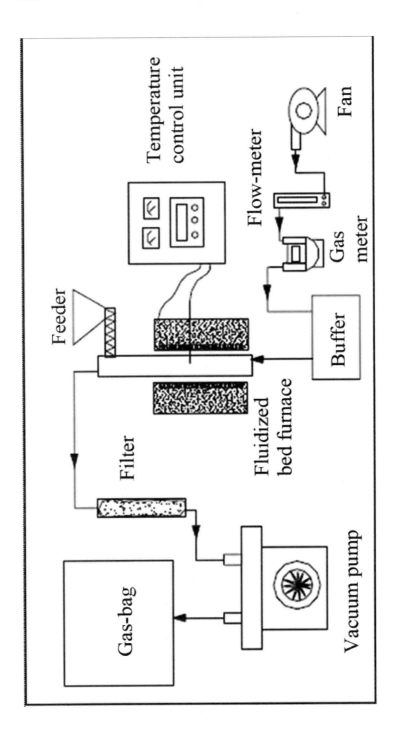

FIGURE 4: Illustration of the Xiao et al. [24] fluidized bed gasifier.

Several types of gasification processes have previously been developed and reported. Their practical performance data, however, have not necessarily been satisfactory for universal application. If char is produced in the gasification, it needs to be further processed and/or burnt. Other gasification schemes (mainly in pilot scale) use a great deal of expensive pure oxygen, whilst others necessitate considerable amounts of expensive materials such as coke and limestone. An ideal gasification process for PSW should produce a high calorific value gas, completely combusted char, produce an easily recoverable ash and should not require any additional installations for air/water pollution abatement.

Early gasification attempts of plastics present in municipal solid waste have been reported since the 1970s [17,18]. The gasification with high calorific value fuel gas obtained from PSW was demonstrated in research stages and results were reported and published in literature for PVC [19], PP [20] and PET [21]. The need for alternative fuels has lead for the co-gasification of PSW with other types of waste, mainly biomass. Pinto et al. [15,22] studied the fluidized bed co-gasification of PE, pine and coal and biomass mixed with PE. Slapak et al. [23] designed a process for steam gasification of PVC in a bubbling fluidized bed.

Xiao et al. [24] co-gasified five typical kinds of organic components (wood, paper, kitchen waste, PE-plastic, and textile) and three representative types of simulated municipal solid waste (MSW) in a fluidized-bed (400°C - 800°C). It was determined that plastic should be gasified at temperatures in excess of 500°C to reach a lower heating value (LHV) of 10,000 kJ/Nm^3. Figure 4 represents the process of co-gasification used in their study.

Gasification produces three different phases: a solid phase (char), a liquid phase (tars) and a gas phase [25]. First products yielded are usually in the range of C_{20} to C_{50}. These products are cracked in the gas phase to obtain lighter hydrocarbons, as ethene and propene, which are unstable at high temperatures and react to form aromatic compounds as benzene or toluene. In thermo-chemical treatment of polyolefins (mainly PE and PP), products obtained mainly depend on cracking reactions in the gas phase. Long residence time of volatiles in reactors and high temperatures decreases tar production but increases char formation [26].

It is believed that increasing temperatures above 500°C and prolonging gas residence time, reduces the tar content in the gas product from gasification of PSW, ASR, MSW and even mixtures of coal, biomass and PSW [22, 27-30].

At temperatures above 800°C, larger paraffins and olefins produced from decomposition of plastics are cracked into H_2, CO, CO_2, CH_4 and lighter hydrocarbons [31].

In PSW gasification, the endothermic gasification reactions involving steam and CO_2 [32,33] (and high heating rates create a char which is more reactive and easier to deal with [34,35]. As a result of these reactions, high gasification temperatures have been reported to increase the H_2 concentration, gas yield [15] and sometimes LHV [36] for a wide range of gasification configurations and oxidizing media.

12.3.2 GASIFICATION OF AUTOMOTIVE SHREDDER RESIDUE (ASR)

A specific and recent development involves the gasification of ASR, where plastics are significantly present.

The study of Harder and Forton [37] describes the process developed by Schwarze Pumpe, producing methanol as a fuel. Sequential gasification and incineration tests were reported by Mancini et al. [38] and make use of a rotary kiln operated between 850°C and 1120°C with an air factor <1. Combustion of the gases is completed in a secondary afterburner chamber. The system includes a boiler (steam at 43 bar, 430°C) and turbine. The capacity was on average 2400 kg/h during the tests. The paper fully describes the characteristics of the ASR used, the combustion properties, the ash analyses, the composition of the exhaust gases, the process operational problems, the residue management, and the energy efficiency. It was concluded that the process requires minor modifications. Atmospheric emissions were invariably considerably below the legal limits.

A similar full report of a sequential gasification and combustion unit to treat ASR, using a fluidized bed gasifier (590°C) followed by a cyclonic afterburner (1400°C) is described by Viganò et al. [39] and by Cho et al. [40]. All operational details and properties of the different process streams are included in the publications. Operation of the afterburner at 1400°C moreover produces a vitrified (and hence inert) slag. It is concluded that the sequential gasification and combustion system reaches appealing energy and environmental performances, despite the unfavourable characteristics of ASR.

A catalytic gasification of ASR with hydrogen generation is presented by Lin et al. [41], using a lab-scale fixed-bed downdraft gasification process. A 15 wt% NiO/Al_2O_3 catalyst is used at 760°C–900°C. It is predicted that such a process, conducted at 46.2 atm would yield 87% of syngas, with a 0.27 m^3 reactor allowing to ultimately produce 100 kW of electricity starting from 220 kg/h of ASR. Further tests are however needed.

Sequential microwave pyrolysis and high temperature agent gasification (HTAG) experiments were performed by Donaj et al. [42,43]. The research suggests using the liquid and solid residue from the microwave pyrolysis as fuel for the HTAG process. In this process a gasifying agent (steam, air or an air/steam mixture) is heated to temperatures above 900°C, providing all the heat needed for gasification.

The use of HTAG can lead to higher conversion of fuel to gas, higher hydrogen yields and lower tar content in comparison to conventional gasification [42,43]. Although the results are fairly supportive for the gasification of the liquid pyrolysis residue, additional process optimization is required towards the pyrolysis of the solid residue. The use of a sequential gasification and combustion system (at very high temperatures) as reported by Mancini et al. [38], by Viganò et al. [39] and by Cho et al. [40] demonstrates that atmospheric emissions were considerably below the legal limits. In the specific Japanese case, the operation of the afterburner at 1400°C moreover produces a vitrified (and hence inert) slag. It is concluded that the sequential gasification and combustion system reaches appealing energy and environmental performances, despite the unfavourable characteristics of ASR.

12.3.3 LARGE-SCALE GASIFICATION TECHNOLOGIES

One of the most common technologies is the Waste Gas Technology UK Limited (WGT) process (Figure 5). Different kinds of wastes (PSW, MSW, sludges, rubbers, wood and straw) are dried and mechanically pre-treated, sorting out incombustibles and granulated to optimum sized particles and fed into a cylindrical reactor for gasification at 700°C–900°C to yield a HCV gas [44]. Upon discharge and subsequent separation of gas and char, the latter may be utilized via combustion in a boiler to raise steam while the gas is quenched and cleaned of contaminants prior to its use in a gas engine or turbine and possibly CCGT applications.

The Texaco gasification process is by far the most common and well known technology. First pilot scale experiments (10 tons/day) were carried out in the US [45]. Figure 6 reviews the process which consists of two parts, a liquefaction step and an entrained bed gasifier. In the liquefaction step the plastic waste is mildly thermally cracked (depolymerisation) into synthetic heavy oil and some condensable and non-condensable gas fractions. The non-condensable gases are reused in the liquefaction as fuel (together with natural gas). Oil and condensed gas produced are injected to the entrained gasifier [46]. The gasification is carried out with oxygen and steam at a temperature of 1200°C–1500°C. After a number of cleaning processes (amongst others, HCl and HF removal), a clean and dry synthesis gas is produced, consisting predominantly of CO and H_2, with smaller amounts of CH_4, CO_2, H_2O and some inert gases [47]. Table 1 summarizes the products from the input criteria and process.

In the case of PSW severely contaminated with other waste products (including contaminated wood, waste water purification sludge, waste derived fuel, paper fractions, etc.), the SVZ process constitutes the optimum solution. The input is fed into a reactor (kiln), together with lignite (in the form of briquettes) and waste oil. Oxygen and steam are used as gasification media, and are supplied in counter flow with the input materials [47]. This processes synthesis gas (a mixture of hydrogen and CO), liquid hydrocarbons, and effluent. The gas is used mainly for methanol production and around 20% is used for electricity production. One of the main advantages of this process is its tolerance for various input parameters. Tukker et al. [47] stated a number of acceptance criteria for the SVZ process, summarized below:

- Particle size: >20 to 80 mm;
- Chlorine content: 2% as default, though higher concentrations are tolerable;
- Ash content: up to 10% or more.

Gasification of Plastic Waste as Waste-to-Energy 253

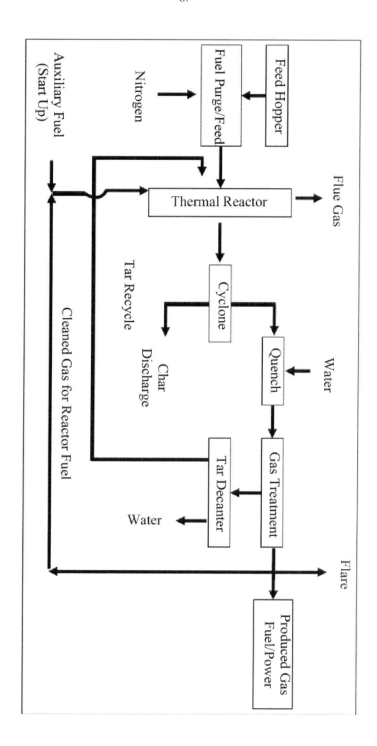

FIGURE 5: WGT process schematic [44].

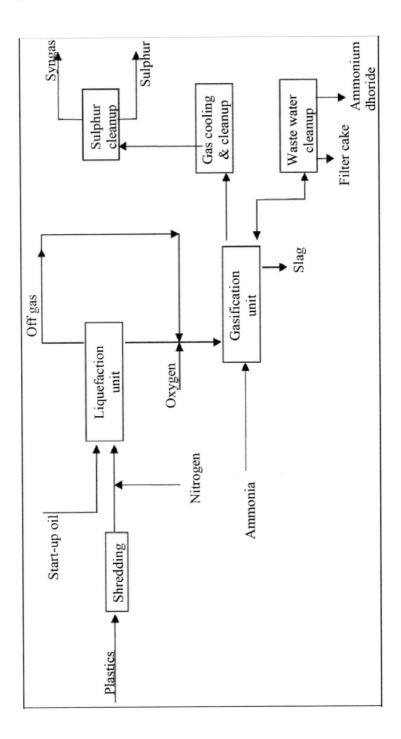

FIGURE 6: Texaco gasification process schematic diagram, showing both stages involved (liquefaction and gasification) [45].

Gasification of Plastic Waste as Waste-to-Energy

TABLE 1: Input criteria and expected output for the Texaco gasification process [47].

	Input Criteria
Feedstock	Dry to the touch, not sticky, shredded or chipped to <10 cm, with <1% under 250 μm
PSW content	>90 wt%
Free metals	<1 wt%
PVC content	<10 wt%
Ash content	<6 wt%
Residual moisture	<5 wt%
Paper content	<10 wt%
	Expected Products
Synthesis gas	350,000 Nm^3/day (predominantly H$_2$/CO) out of 150 tons of PSW/day
Vitrified slag	–
Fines	High quality, equivalent to fly ash

As a producer of chlorine and vinylchloride, Akzo Nobel started a process for mixed PSW gasification. The process consists of two separate circulating fluid bed (CFB) reactors at atmospheric pressure (Figure 7). The first is a gasification reactor in which waste (usually rich with PVC) is converted at 700°C–900°C into product gas (fuel and HCl gas) and tars. The second unit is a combustion reactor that burns the residual tar to provide heat for the gasification process. Circulating sand between the gasifier and combustor transfers heat between the two reactors. Both reactors are of the riser type with a very short residence time. This type of reactor allows a high PVC waste throughput. If the input contains a lot of PE and PP, the output will contain a significant amount of propylene and ethylene [47].

12.4 EXPERIMENTAL WORK

12.4.1 FEEDSTOCK MATERIAL

As PSW samples, PET, PE, PP and PS were selected. They were either chopped into 1–2 cm long chips with maximum thickness below 0.5 mm, or used as pellets of about 3 to 5 mm.

12.4.2 FLUIDIZED BED GASIFICATION REACTOR

The gasification experiments were performed in a bubbling fluidized reactor. The experimental set-up is illustrated in Figure 8.

The bubbling fluidized bed is manufactured from Incoloy steel, and externally insulated. The air flow rate was controlled by flow meter. The heat-input from the electrical heater (50 kW) was set by the bed temperature. Vapours, gases and fluidization gas were exhausted by an induced draft fan, passing a water-cooled condenser and a liquid collection tank. Life steam (110°C, 5 bar) could be added in the windbox of the fluidized bed, i.e. below the distributor plate, built as a sintered metal plate with pressure drop of about 3000 Pa at the operating flow rate and temperature.

The air flow rate was maintained at about 20% to 25% of the stoichiometric air flow needed for the combustion of each specific PSW material (Equivalence Factor, EF, 20% - 25%), and the velocity in the bed varied from 2 to 4 times the minimum fluidisation velocity, U_{mf}. For some of the experiments, a steam dilution was also applied, reducing EF to below 15%.

A known weight (between 0.3 and 0.5 kg) of the PSW sample was placed into a perforated basket. The perforations enabled the fluidized bed sand to fluidize within the voidage of the PSW chips.

The fluidized bed consisted of about 15 kg of quartz sand (150 - 300 μm, 2600 kg/m^3), for a total static bed height of about 35 cm.

Gases were exhausted to the atmosphere, after appropriate analysis of CO, CO_2, H_2 and CH_4. Temperatures between 550°C and 800°C were tested. In additional experiments, live-steam (110°C, 5 bar) was directly injected directly into the fluidization air flow, thus further reducing the partial pressure of O_2 being present. All experiments were twice repeated under identical operating conditions, and average results were used to determine the kinetics of the PSW gasification.

The progress of the gasification reaction was monitored through the on-line measurement of the CO production. A first order gasification reaction was assumed to be valid, allowing the calculation of the reaction rate constants from PSW conversion as measured by CO production. The order of the reaction was determined by Brems et al. [7], and indeed found to be close to 1 for all polymers.

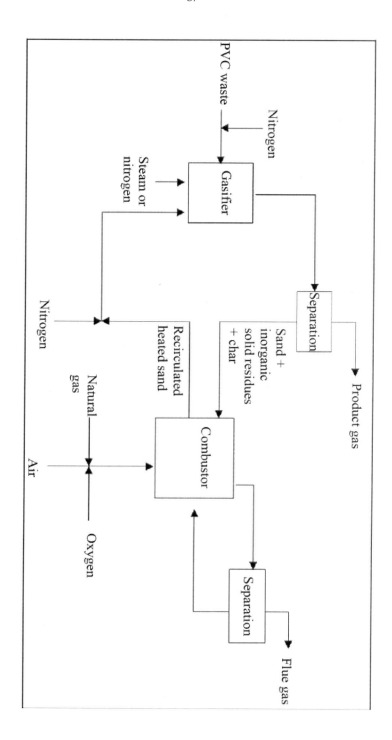

FIGURE 7: Akzo Nobel process schematic diagram [47].

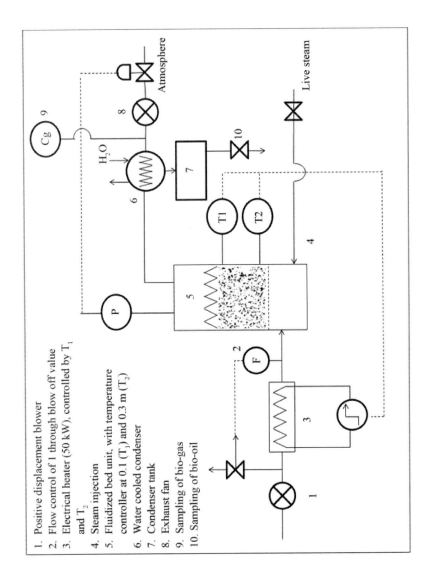

1. Positive displacement blower
2. Flow control of 1 through blow off value
3. Electrical heater (50 kW), controlled by T_1 and T_2
4. Steam injection
5. Fluidized bed unit, with temperature controller at 0.1 (T_1) and 0.3 m (T_2)
6. Water cooled condenser
7. Condenser tank
8. Exhaust fan
9. Sampling of bio-gas
10. Sampling of bio-oil

FIGURE 8: Schematic diagram of the gasification set-up

12.4.3 EXPERIMENTAL RESULTS

The optimum temperatures of gasification (max. gas yield) are defined in Table 2, together with the activation energy, E_a, and the pre-exponential factor, A, of the Arrhenius kinetic expression.

The reaction rate constant, k, can be determined using the Arrhenius equation, reproduced below:

$$k = Ae^{-E_a/RT} \tag{1}$$

Using the experimental data, the reaction rate constant for the different PSW materials can be predicted at any temperature.

At the respective optimum T of PE, k equals 5.35 s^{-1}, representing a fast reaction rate.

The required residence time of PE in the reactor can be determined from the overall reaction kinetics, as given for a first order reaction:

$$-dW/dt = kW \tag{2}$$

Integration of Eq.2 and introducing X_p, the degree of conversion (the weight percentage of initial solid that is successfully converted into the product), yields:

$$-\ln(1 - X_p) = kt \tag{3}$$

From experimentation it was seen that the gasification of PSW is very efficient and a conversion factor, X_p, of 0.99 is easily achieved.

The required residence time, t, can then be evaluated as:

$$-\ln(1 - 0.99) = 5.35t \tag{4}$$

For the case of PE, this leads to a required residence time, t, of 0.86 s. Gasification is hence a very fast reaction at the optimum temperatures, and ideally suited for a fluidised bed gasifier.

TABLE 2: Experimental values of T, Ea and A.

PSW	T(K)	A (s^{-1})	E_a (kJ/mol)
PS	953	2.94×10^{14}	212
PET	968	2.94×10^{16}	238
PE	1008	5.12×10^{15}	289
PP	1018	1.99×10^{13}	187

12.5 CONCLUSIONS

From the previous review and experimental work, the gasification of PSW can certainly be developed into a valid recycling route for PSW, producing a syngas, rich in H_2 and CO. Although references of industrial scale application are given in the literature, the future breakthrough of the process will require further experimental work to improve the equipment design and product optimisation. Advances in that area will aid in the improvement and more widespread use of gasification reactors.

REFERENCES

1. Baeyens, J., Brems, A. and Dewil, R., (2010) Recovery and recycling of post-consumer waste materials—Part 2. Target wastes (glass beverage bottles, plastics, scrap metal and steel cans, end-of-life tyres, batteries and household hazardous waste). International Journal of Sustainable Engineering, 3, 232-245. doi:10.1080/19397038.2010.507885
2. Al-Salem, S.M., Lettieri, P. and Baeyens, J. (2009) Recycling and recovery routes of plastic solid waste (PSW): A review. Waste Management, 29, 2625-2643. doi:10.1016/j.wasman.2009.06.004
3. Al-Salem, S.M., Lettieri, P. and Baeyens, J., (2010) The valorization of plastic solid waste (PSW) by primary to quaternary routes: From re-use to energy and chemicals. Progress in Energy and Combustion Science, 36, 103-129. doi:10.1016/j.pecs.2009.09.001

4. Ahrenfeldt, J. (2007) Characterisation of biomass producer gas as fuel for stationary gas engines in combined heat and power production. Ph.D. Thesis, Technical University of Denmark, Lyngby.
5. Yoshioka, T., Gause, G., Eger, C., Kaminsky, W. and Okuwaki, A. (2004) Pyrolysis of polyethylene terephthalate in a fluidised bed plant. Polymer Degradation and Stability, 86, 499-504. doi:10.1016/j.polymdegradstab.2004.06.001
6. Smolders, K. and Baeyens, J. (2004) Thermal degradation of PMMA in fluidised beds. Waste Management, 24, 849-857. doi:10.1016/j.wasman.2004.06.002
7. Brems, A., Baeyens, J., Beerlandt, J. and Dewil, R. (2011) Thermogravimetric pyrolysis of waste polyethylene-terephthalate and polystyrene: A critical assessment of kinetics modeling. Resources, Conservation and Recycling, 55, 772-781. doi:10.1016/j.resconrec.2011.03.003
8. Steiner, C., Kameda, O., Oshita, T. and Sato, T. (2002) EBARA's fluidized bed gasification: Atmospheric 2 × 225 t/d for shredding residues recycling and two-stage pressurized 30 t/d for ammonia synthesis from waste plastics. Proceedings of Second International Symposium on Feedstock Recycle of Plastics and Other Innovative Plastics Recycling Techniques, Ostend, 8-11 September 2002.
9. Aguado, J., Serrano, D.P., Miguel, G.S., Escola, J.M. and Rodriguez, J.M. (2007) Catalytic activity of zeolitic and mesostructured catalysts in the cracking of pure and waste polyolefins. Journal of Analytical and Applied Pyrolysis, 78, 153-161. doi:10.1016/j.jaap.2006.06.004
10. Mastellone, M.L. (1999) Thermal treatments of plastic wastes by means of fluidised bed reactors. Ph.D. Thesis, Second University of Naples, Naples.
11. Arena, U. and Mastellone, M.L., (2006) Fluidized bed pyrolysis of plastic wastes. In: Scheirs, J. and Kaminsky, W., Eds., Feedstock Recycling and Pyrolysis of Plastic Wastes: Converting Waste Plastics into Diesel and Other Fuels, John Wiley & Sons, Chichester. doi:10.1002/0470021543.ch16
12. Scheirs, J. (1998) Polymer recycling. Wiley, New York.
13. Vermeulen, I., Van Caneghem, J., Block, C., Baeyens, J. and Vandecasteele, C. (2011) Automotive shredder residue (ASR): Reviewing its productions from end-of-life vehicles (ELVs) and its recycling, energy and chemicals valorization. Journal of Hazardous Materials, 190, 8-27. doi:10.1016/j.jhazmat.2011.02.088
14. Wallmann, P.H., Thorsness, C.B. and Winter, J.D. (1998) Hydrogen production from wastes. Energy, 23, 271-278. doi:10.1016/S0360-5442(97)00089-3
15. Pinto, F., Franco, C., Andre, R.N., Miranda, M., Gulyurtlu, I. and Cabrita, I. (2002) Co-gasification study of biomass mixed with plastic wastes, Fuel, 81, 291-297. doi:10.1016/S0016-2361(01)00164-8
16. VTT (2004) Power production from waste and biomass IV. Proceedings of the VTT Symposium, Finland, 8-10 April 2002.
17. Buekens, A.G. (1978) Resource recovery and waste treatment in Japan. Resource Recovery and Conservation, 3, 275-306. doi:10.1016/0304-3967(78)90011-2
18. Hasegawa, M., Fukuda, X. and Kunii, D. (1974) Gasification of solid waste in a fluidized bed with circulating sand. Conservation and Recycling, 3, 143-153. doi:10.1016/0361-3658(79)90004-3
19. Borgianni, C., Filippis, P.D., Pochetti, F. and Paolucci, M. (2002) Gasification process of wastes containing PVC. Fuel, 14, 1872-1833.

20. Xiao, G., Jin, B., Zhou, H., Zhong, Z. and Zhang, M. (2007) Air gasification of polypropylene plastic waste in fluidized bed gasifier. Energy Conversion and Management, 48, 778-786. doi:10.1016/j.enconman.2006.09.004
21. Matsunami, J., Yoshida, S., Yokota, O., Neuzka, M., Tsuji, M. and Tamaura, Y. (1999) Gasification of waste tyre and plastic (PET) by solar thermochemical process for solar energy utilization. Solar Energy, 65, 21-23. doi:10.1016/S0038-092X(98)00085-1
22. Pinto, F., Franco, C., Andre, R.N., Tavares, C., Dias, M. and Gulyurtlu, I. (2003) Effect of experimental conditions on co-gasification of coal, biomass and plastic wastes with air/steam mixture in a fluidized bed system. Fuel, 82, 1967-1976. doi:10.1016/S0016-2361(03)00160-1
23. Slapak, M.J.P., Kasteren, J.M.N.V. and Drinkenburg, A.A.H. (2000) Design of a process fors team gasification of PVC waste. Resources, Conservation and Recycling, 30, 81-93. doi:10.1016/S0921-3449(00)00047-1
24. Xiao, G., Ni, M., Chi, Y., Jin, B., Xiao, R., Zhong, Z. and Huang, Y. (2009) Gasification characteristics of MSW and ANN prediction model. Waste Management, 29, 240- 244. doi:10.1016/j.wasman.2008.02.022
25. Aznar, M.P., Caballero, M.A., Sancho, J.A. and Francs, E. (2006) Plastic waste elimination by co-gasification with coal and biomass in fluidized bed with air in pilot plant. Fuel Processing Technology, 87, 409-420. doi:10.1016/j.fuproc.2005.09.006
26. Cozzani, V., Nicolella, C., Rovatti, M. and Tognotti, L., (1997) Influence of gas phase reactions on the product yields obtained in the pyrolysis of polyethylene. Industrial and Engineering Chemistry Research, 36, 342-348. doi:10.1021/ie950779z
27. Stiles, H.N. and Kandiyoti, R. (1989) Secondary reactions of flash pyrolysis tars measured in a fluidized bed pyrolysis reactor with some novel design features. Fuel, 86, 275-282. doi:10.1016/0016-2361(89)90087-2
28. Zolezzi, M., Nicolella, C., Ferrara, S., Iacobucci, C. and Rovatti, M. (2004) Conventional and fast pyrolysis of automotive shredder residues (ASR). Waste Management, 24, 691-699. doi:10.1016/j.wasman.2003.12.005
29. Miscolczi, N., Bartha, L., Deák, G. and Jóver, B. (2004) Thermal degradation of municipal solid waste for production of fuel-like hydrocarbons. Polymer Degradation and Stability, 86, 357-366.
30. Ciliz, N.K., Ekinci, E. and Snape, C.E. (2004) Pyrolysis of virgin and waste polyethylene and its mixtures with waste polyethylene and polystyrene. Waste Management, 2, 173-181. doi:10.1016/j.wasman.2003.06.002
31. Ponzio, A., Kalisz, S. and Blasiak, W. (2006) Effect of operating conditions on tar and gas composition in high temperature air/steam gasification (HTAG) of plastic containing waste. Fuel Processing Technology, 3, 223-233. doi:10.1016/j.fuproc.2005.08.002
32. Franco, C., Pinto, F., Gulyurtlu, I. and Cabrita, I. (2003) The study of reactions influencing the biomass gasification process. Fuel, 82, 835-842. doi:10.1016/S0016-2361(02)00313-7
33. Marquez-Montesinos, F., Cordero, T., Rodriguez-Mirasol, J. and Rodriguez, J.J. (2002) CO_2 and steam gasification of grapefruit skin char. Fuel, 81, 423-429. doi:10.1016/S0016-2361(01)00174-0

34. Zanzi, R., Sjöström, K. and Björnbom, E. (1996) Rapid high-temperature pyrolysis of biomass in a free fall reactor. Fuel, 75, 545-550. doi:10.1016/0016-2361(95)00304-5
35. Zanzi, R., Sjöström, K. and Björnbom, E. (2002) Rapid pyrolysis of agricultural residues at high temperature. Biomass and Bioenergy, 23, 357-366. doi:10.1016/S0961-9534(02)00061-2
36. Narvaez, A., Orio A., Aznar, M.P. and Corella, J. (1996) Biomass gasification with air in an atmospheric bubbling fluidized bed: effect of six operational parameters. Industrial and Engineering Chemistry Research, 35, 2110- 2120. doi:10.1021/ie9507540
37. Harder, M.K. and Forton, O.T. (2007) A critical review of developments in the pyrolysis of automotive shredder residue. Journal of Analytical and Applied Pyrolysis, 79, 387-394. doi:10.1016/j.jaap.2006.12.015
38. Mancini, G., Tamma, R. and Viotti, P. (2010) Thermal process of fluff: Preliminary test on a full scale treatment plant. Waste Management, 30, 1670-1682. doi:10.1016/j.wasman.2010.01.037
39. Vigano, F., Consonni, S., Grosso, M. and Rigamonti, L., (2010) Material and energy recovery from automotive shredder residue (ASR) via sequential gasification and combustion. Waste Management, 30, 145-153. doi:10.1016/j.wasman.2009.06.009
40. Cho, S.J., Jung, H.Y., Seo, Y.C. and Kim, W.H. (2010) Studies on gasification and melting characteristics of automotive shredder residue. Environmental Engineering Science, 27, 577-586. doi:10.1089/ees.2009.0389
41. Lin, K.S., Chowdhury, S. and Wang, Z.P. (2010) Catalytic gasification of automotive shredder residues with hydrogen generation. Journal of Power Sources, 195, 6016-6023. doi:10.1016/j.jpowsour.2010.03.084
42. Donaj, P., Yang, W., Blasiak, W. and Forsgren, C. (2010) Recycling of automotive shredder residue with a microwave pyrolysis combined with high temperature steam gasification, Journal of Hazardous Materials, 182, 80-89. doi:10.1016/j.jhazmat.2010.05.140
43. Donaj, P., Blasiak, W., Yang, W. and Forsgren, C. (2011) Conversion of microwave pyrolysed ASR's char using high temperature agents. Journal of Hazardous Materials, 185, 472-481. doi:10.1016/j.jhazmat.2010.09.056
44. WGT (2002) Waste gas technology energy from waste. http://www.wgtuk.com/ukindex.html
45. Weissman, R. (1997) Recycling of mixed plastic waste by the Texaco gasification process. In: Hoyle, W. and Karsa, D.R., Eds., Chemical Aspects of Plastics Recycling, The Royal Society of Chemistry Information Services, Cambridge.
46. Croezen, H. and Sas, H. (1997) Evaluation of the Texaco gasification process for treatment of mixed household waste. Final Report of Phase 1 and 2, CE, Delft, The Netherlands.
47. Tukker, A., de Groot, H., Simons, L. and Wiegersma, S. (1999) Chemical recycling of plastic waste: PVC and other resins. European Commission, DG III, Final Report, STB-99-55 Final, Delft, The Netherlands.

Author Notes

CHAPTER 2

Competing Interests
The authors declare that they have no competing interests.

Author Contributions
AHP carried out the country-specific research, collected the questionnaires, and wrote the manuscript, tables, and figures. HO conducted the previous applied studies in Japan and collected the other previous studies using similar methodology for the references. NK developed the methodology. All authors read and approved the final manuscript.

Author Information
AHP has completed her Master of Engineering degree form Trier University of Applied Sciences, Environmental Campus Birkenfeld - Germany. She is currently enrolled in the Ph.D. programme of Waseda University, Graduate School of Environment and Energy Engineering - Japan. OH is an associate professor and NK is a professor in Waseda University with expertise in Life Cycle Assessment of products, technologies, services, and social-economical systems.

Acknowledgments
The authors thank the Japan Ministry of Education, Culture, Sports, Science and Technology for the financial support throughout this research.

CHAPTER 3

Competing Interests
The authors declare that they have no competing interests.

Author Contributions

All authors (LM, DK, ML and PC) contributed jointly to all aspects of the work reported in the manuscript. All authors have read and approved the final manuscript.

Acknowledgments

This work is based on the research that was carried out in the framework of a LIFE+project entitled: Development and demonstration of an innovative method of converting waste into bioethanol, Waste2Bio, (LIFE 11 ENV/GR/000949, 2012–2015), which is co-financed by the European Commission. Paul Christakopoulos thanks Bio4Energy, a strategic research environment appointed by the Swedish government, for supporting this work.

CHAPTER 4

Acknowledgments

The authors acknowledge the Biomass funding from the Ability R&D Energy Research Centre (AERC) at the School of Energy and Environment in the City University of Hong Kong. We are also grateful to the donation from the Coffee Concept (Hong Kong) Ltd. For the 'Care for Our Planet' campaign, as well as a grant from the City University of Hong Kong (Project No. 7200248). Authors acknowledge the Industrial Technology Funding from the Innovation and Technology Commission (ITS/323/11) in Hong Kong.

Conflicts of Interest

The authors declare no conflict of interest.

CHAPTER 5

Acknowledgments

This work was financially supported by the Polish Ministry of Science and Higher Education, project no. 2P04G 05928.

CHAPTER 6

Acknowledgments

The authors would like to acknowledge the cooperation of the engineers as well as the managers at the Isfahan "composting factory" for providing the tour and description of the steps and stages during the operations at the facility, to the senior author few years ago. The authors would also like to acknowledge the efforts of the chief engineer and staff for providing the pictures and related information that are being presented here and at related conferences.

CHAPTER 8

Conflicts of Interest

The authors declare no conflict of interest.

CHAPTER 9

Competing Interests

The authors declare that they have no competing interests.

Author Contributions

RL performed all practical experiments and prepared article manuscript; JO and NS made critical review of the manuscript, approved submission of article for publication. All authors read and approved the final manuscript.

Acknowledgments

The authors wish to acknowledge the Engineering and Physical Sciences Research Council, UK that finances this research through its SUPERGEN-Bioenergy research consortium.

CHAPTER 10

Acknowledgments

We would like to acknowledge the National Science Council of Taiwan for providing a financial support on this research.

CHAPTER 11

Conflicts of Interest

The authors declare no conflict of interest.

CHAPTER 12

Acknowledgments

The present research was supported by the Fundamental Research Funds for the Central Universities (Program RC1101).

Index

2,5-dimethylfuran (DMF), xx–xxii,
xxv, 5, 7–12, 14–22, 24–28, 35–36,
38–42, 44–45, 47, 51, 55–56, 58–62,
64, 66–67, 73, 75–78, 80–84, 86,
88–89, 91, 94, 98–99, 105, 107–110,
112–114, 118, 120, 126–129, 132,
134–135, 139, 143–144, 147–150,
153–155, 157–158, 161, 163–164,
167, 172–174, 176–178, 184,
187–189, 191–192, 194–201, 204,
206, 214–215, 217–222, 226–228,
234–235, 238, 240, 242–245,
250–252, 255–256, 259–263, 265

A

acid rain, xx, 12–13, 15, 22, 25–27
activation energy (E_a), xxvi, 129, 134,
192, 195, 259–260
aeration, 99
air-fuel ratio, xxvi, 233–235
alcohol, xxiii, 82, 105, 107, 121–125,
127–129, 134–137, 139–140
aldol-condensation, 105, 117–118,
137, 139
algae, xvii, 103–104
alternator, 156, 164
anaerobic digestion (AD), xviii, xx,
5–8, 10, 12, 16, 18, 20–21, 25–27,
29, 33, 48, 52, 86, 166, 188, 195,
197, 208
analytic hierarchical process (AHP),
xx, 14, 22–23, 26, 265
animal feed, 36, 172

ash, 5, 12, 19, 30, 38, 48, 158, 176,
184, 194, 218, 220, 223, 226,
228–229, 246, 249–250, 252, 255
Aspen Plus, xxv, 215, 220, 224–229,
232, 238–240
Aspergillus awamori, 56–57, 59, 66

B

bacteria, 53, 77, 83–85
bio-ethanol, xvii, 35
biocoal, xxiv, 190–191, 197–199, 201,
204
biodiesel, 3, 108
biofuels, xvii–xviii, 3–4, 35–36, 41,
50–52, 103–105, 107, 109, 111, 113,
115, 117, 119, 121, 123, 125, 127,
129, 131, 133–135, 137, 139–141,
143, 214
  biofuel yield, 41
  first-generation biofuels, xvii, 35
  liquid biofuels, 3
  second-generation biofuels, xvii, 35
biogas, 3–4, 16, 20–21, 25, 27, 30, 36,
38, 41, 51–52, 153, 163
biological oxygen demand (BOD),
12–13, 38
biomass
  aquatic biomass, 104
  biomass fraction, 186
  biomass fractions, 184
  biomass moisture, 214
  biomass pretreatment, 191, 208
  biomass torrefaction, 191

lignocellulosic, xxiii, 36–37, 51, 53, 103, 106, 108–110, 112–113
biorefineries, 55
biostabilization, 148, 153–154
biotransformation, xxii, 74–75, 77, 83–84
bitumen, 74
butanol, 108, 134, 136, 139

C

capital cost, 4, 6
carbohydrates, 44, 53, 55, 107, 141, 143
carbon, xxi–xxii, xxiv–xxv, 3–4, 28, 55–57, 76, 78, 81, 83–84, 98, 105, 107, 110, 114, 117, 123, 130, 172, 175–176, 184, 186–187, 190–191, 204, 208, 214, 218–219, 226, 239–240, 246
   carbon credits, 3, 28
   carbon dioxide (CO_2), xvii, xxv, 3–5, 12–13, 15–16, 21, 26–27, 30, 103, 107, 172–173, 175–177, 200, 204, 207, 218–221, 234–235, 238, 240, 243, 250, 252, 256, 262
   carbon monoxide (CO), xviii–xix, xxiv–xxv, 3, 13, 15, 26–27, 30, 41, 52–53, 84–86, 125, 131, 133, 140, 147, 156–157, 164–166, 172, 175–176, 180–184, 186, 188–191, 195, 197–199, 201, 204–208, 214, 218–221, 234–235, 238, 243, 246, 249–250, 252, 255–256, 260–262, 266
   carbon sequestration, 4
carbonization, xviii, xxiv–xxv, 145, 189–195, 197–199, 201, 203–207, 209
carcinogens, 199, 201

catalyst, xxiii, 44, 82, 105, 112, 114, 117–118, 121–122, 125, 127–135, 138–139, 143, 175, 186, 243, 251, 261
   metal catalysts, xxiii, 121–122, 134
catalytic cracking, 242–243
cell membranes, 82
cellulose, xxi, 36–41, 44–45, 48, 107–108, 110, 112
chemical oxygen demand (COD), 12–13, 38
chlorofluorocarbon, 12
cleavage, 55, 110, 112, 127, 134–135
co-generation, 16, 25, 147, 157, 164–165
coal, xxiv–xxvi, 12–13, 15–16, 21, 25–27, 30, 85, 172, 175, 184, 188, 190–191, 197–199, 201, 204, 206–208, 225, 228, 239–240, 243, 249–250, 262
   coal mining, 198
   coal substitution, 197–198, 201, 204, 206
coke, 108, 132–133, 249
combustion, xvii–xviii, xxiv, xxvi, 3–5, 103, 149, 154–155, 157–158, 169, 171–177, 179–188, 190, 192, 207, 214–215, 218, 220–222, 228–229, 232, 239, 246, 250–252, 255–256, 260, 263
composting, xviii, xx–xxiii, 4, 7–8, 10, 16–20, 26, 28–29, 39, 71, 73–78, 80–86, 94–96, 99–100, 164, 172, 267
   mechanical composting, 94
   windrow composting, 8, 10, 20
consumption, xx, xxiii, 13–16, 21–22, 25–29, 89, 127, 148, 168, 177, 198, 202, 205–206, 242
conversion efficiency, 4–5, 220, 238
corn, 35–36, 41, 50, 52, 104, 107, 188, 215, 239

Index

corn stover, 41, 50, 52, 188, 215, 239
cost, xxiv, 4–6, 36, 51, 88, 94, 148, 154, 165–166, 168, 190, 205, 207–208, 246
    management costs, 154

D

deactivation rate, 60, 133
decarbonylation, xxiii, 122, 125, 129–132, 134, 136, 139
decomposition, xxi–xxii, 74–75, 77, 82–85, 131, 205, 226–229, 250
dehydration, 110, 112, 114–115, 117
diesel, 21, 104–105, 117, 137, 261
dinitro-3,5-salicilic acid (DNS), 47
dolomite, 219
dried distillers' grains with solubles (DDGS), 172–177, 180–184
dry material (DM), xxi, 35, 37, 39–47, 49
drying, 191–192, 195, 199, 201, 205–208, 219, 226, 228–229

E

EcoInvent, 11, 14, 16, 20, 30, 198
economic development, 8
ecotoxicity, xxv, 199, 201
education, 37, 89, 94, 265–266
efficiency, 4–5, 40–41, 44, 62, 86, 107, 112, 154–156, 171, 174, 190, 192, 194–195, 214–215, 220, 235, 238, 246, 250
    conversion efficiency, 4–5, 220, 238
electricity, xviii, xx, xxiv, 3, 15–16, 20–21, 25–27, 30, 157, 171, 190, 197–199, 202, 205–206, 213–214, 251–252
    electricity demand, xxiv, 171
energy
    energy depletion, xx, 12–13, 15, 22, 25–27
    energy efficiency, 107, 250
    energy prices, 4
    energy recovery, xx, xxiii–xxvi, 3–7, 9–13, 15–21, 23, 25, 27–29, 31, 148–149, 153, 163–166, 168, 172, 195, 197, 206, 215, 242, 263
    thermal energy, 156, 165
environmental
    environmental assessment, 154, 167, 209
    environmental impact, xx, xxiv–xxv, 10, 15, 28, 148, 164, 172
    environmental load factor (ELF), 13–14, 22, 25
    environmental load point (ELP), xx, 7, 10–15, 22, 24, 28–30
equilibrium, 127, 133, 225, 229, 232
equivalence ratio (ER), 191–192, 214
esterification, 108
esters, 74, 76–77, 86
ethanol, xvii, xix–xxii, 3, 35–53, 56, 70, 76, 105, 136, 140, 172, 188
    ethanol production efficiency, 40–41, 44
    ethanol recovery, 37
etherification, xxiii, 117–118
eutrophication, xxv, 200–201

F

fatty acids, xxii, 74–76, 78, 80–86
    fatty acid carbon (FAC), xxii, 76–78, 82–83

long-chain fatty acids (LCFAs), 82, 86
fatty acids methyl esters (FAMEs), 76–77, 83
feasibility, xix, xxi, xxiii, 5–6, 51, 56, 68, 88, 140, 190
feedlots, 4
feedstock, xvii–xix, xxvi, 3, 5, 37, 55, 68, 103–104, 110, 139–140, 168, 214, 218, 242–243, 246, 255, 261
    feedstock preparation, 218
fermentation, xxi, 37–41, 44–53, 56–57, 59, 66, 69–70, 95–96, 99, 108, 172, 207
    solid state fermentation (SSF), xxi, 40, 56–57, 59, 66, 69
fertilizer, xviii, xx, 10, 15–16, 20–21, 25–28, 36, 94
Fischer-Tropsch, 108
flame ionization detector (FID), 77
flue gas, 19, 173, 177, 187, 215
fluidized bed, xxiv–xxv, 173–177, 184–188, 208, 213–215, 217–227, 229–231, 233, 235, 237–240, 246, 248–250, 256, 260–263
    circulating fluidized bed (CFB), 188, 214, 239–240, 246, 255
    dual fluidized bed (DFB), 215, 239
food waste, xviii, xxi, 18, 35, 37, 39–45, 47, 49–53, 55–56, 59–62, 64–70
    food waste hydrolysis, xxi, 56, 60–62, 67
forestry, 213
formic acid, 110
fractionation, 205
free radicals, 175–176, 184, 186
freeboard, 175, 218, 220, 222, 224, 227, 232
fuel
    fuel cells, 3

fuel fraction, 174, 187
functional unit, 15, 195
fungi, 82–83, 85
furanyl ring, 121–124, 128, 132, 134–135
furfural (FAL), xviii, xxiii, 37, 103, 105, 107, 109–119, 121–139, 141–143

G

garbage, 92, 95, 98
gas chromatograph (GC), xxii, 76–77, 80
gas collection, xx, 10, 16, 19, 27–28
gasification, xviii, xxv–xxvi, 4, 6, 18, 26, 108, 110, 166, 208, 211, 213–235, 237–243, 245–247, 249–263
    high temperature agent gasification (HTAG), 251, 262
gasoline, xix, 104–105, 134, 136–137, 140, 243
global warming, xx, 5, 12–13, 15, 22, 25–27, 200–201, 204, 208
    global warming potential, 25–26
glucoamylase (GA), xxi, 41, 55–69, 84
glucose, xxi, 38, 55–57, 59, 61–65, 67–69, 107, 114–115
greenhouse gas (GHG), xxiv–xxv, 3, 5–6, 21, 31, 36, 148, 162, 164, 190, 204–206, 213

H

halogens, 4
heat, xviii, xxiv, xxvi, 15–16, 21, 60, 104, 127, 131, 149, 154–156, 158, 164, 166, 168–169, 171, 191–192, 194, 197–198, 201, 206, 214–215, 218–220, 228, 239, 242–243, 246, 251, 255–256, 261

Index

heat exchanger, 15, 154, 156, 158, 164, 191–192, 194
heat recovery, 15, 197, 206
heat treatment, 60
heat waste, 15
heating, xxiii–xxiv, 15, 104–105, 110, 149, 157–158, 163, 165–166, 171, 194, 198, 215, 249–250
   heating rate, 110
heavy metals, 4–5, 10
height, 187, 222, 256
household food wastes (HFW), xx, 36–40, 42, 44–49
humidity, 74, 77–78
humification, 74
humin, 74
hydrodynamics, 222–225
hydrogen, xviii, xxiv, 3, 108, 112, 114, 117, 121–123, 126–128, 131, 174–177, 184, 186, 214, 219, 242, 246, 251–252, 261, 263
   hydrogen production, 3, 246, 261
hydrogenation, xxiii, 117, 121–129, 131–132, 134–137, 139, 243
hydrolysis, xxi, 36, 38, 40, 44, 50–52, 56, 59–62, 64, 66–69, 108, 112
hydrophobic, xxi–xxii, 73–79, 81, 83–85
   hydrophobic substances carbon (HSC), xxii, 76–78
hydroxymethyl furfural (HMF), xxiii, 105, 110, 113–114, 118, 139

I

ideal gas law, 227
incineration, xviii, xx, xxiii–xxvi, 3, 6, 8, 10, 12, 16–20, 25–26, 29–30, 55, 145, 147–149, 151, 153–155, 157–159, 161, 163–169, 188, 190–191, 195, 197–198, 201, 204–206, 242, 246, 250

   co-incineration, xxv, 190–191, 195, 197–198, 201, 204–206
   mono-incineration, xxv, 191, 195, 197–198, 201, 204–206
industrialization, 7, 87, 165
insolubility, 75
integrated solid waste management (ISWM), xix, 29, 88, 100, 147–149, 153, 162–163, 165–166
inventory data, 11, 15–16, 19, 197–198
ionic liquid, 114

L

landfill, xx, xxiii–xxv, 4–6, 8, 10, 12, 16–21, 25–28, 36, 55, 85, 88, 99, 148, 153–154, 158, 163–165, 172, 189–191, 195, 197–198, 201, 204–206, 241
   landfill closure, 163
   landfill gas, xx, 5–6, 10, 16, 19–20, 25, 27–28
   sanitary landfill, xx, 8, 10, 16–19, 27–28, 197
leachate, 19–20, 51
life cycle assessment (LCA), xxv, 10–11, 13, 15, 28–30, 148, 167, 190–191, 195, 198, 206–208, 265
life cycle cost (LCC), 190
life style, xxii–xxiii, 88–89
lignin, 38, 48, 53, 82, 85, 107–108, 112, 141
limestone, 174, 176–177, 186–187, 249
lipid, xxii, 74–75, 80, 82–86
liquefaction, xxi, 37, 39–42, 44, 47, 49, 51, 53, 108, 243, 252, 254
liquid fuel replacement, 213
livestock, xvii, 4, 104
lower heating value (LHV), 249–250

M

mass spectrometer (MS), 30, 76, 156
materials recycling facility residue (MRFR), 172–177, 180–184, 187
MatLab, 224–225
mesophilic, 78, 84–85, 208
methane (CH_4), xviii, 3, 5–6, 12–13, 15–16, 19–21, 25–26, 52, 197–198, 204, 207, 214, 218–220, 234–235, 238, 250, 252, 256
moisture, xxii, xxv, 7, 26, 37–38, 48, 76, 86, 184, 192, 194–195, 205, 214, 218, 226, 228–229, 255
monolignols, 107
municipal solid waste (MSW), xvii–xviii, xx–xxv, 4, 6–7, 11, 14–16, 18, 29–30, 51, 73–75, 78, 84–86, 100, 147–151, 153–155, 157–161, 163, 165–169, 171–173, 186, 189–191, 195, 197–198, 205–206, 249–251, 262
  municipal solid waste management (MSWM), 7, 15, 19, 29, 167–168
municipal waste, xvii, xx, 7–8, 10, 12, 16, 20, 26–27, 30, 86–87, 89, 91, 93, 95, 97, 99, 168
  municipal waste treatment, 8, 10

N

natural gas, 12–13, 15–16, 21, 25, 246, 252
natural resources, 89, 242
nitrogen, xxiv, 56–57, 59, 68, 86, 174, 176, 178, 184, 204, 218–219, 240
nylon, 243

O

oil, xvii, 12–13, 15–16, 21, 27, 49, 84–85, 104, 108, 110, 118, 142, 197, 215, 219–220, 225, 242, 252

open dumping, 8, 10
oxidation, 85, 175–176, 186, 209, 214–215, 235, 246
oxygen, xvii–xviii, xxv–xxvi, 12, 20, 38, 82, 103, 107–108, 110, 112, 117, 174–176, 184, 191, 214, 226, 240, 246, 249, 252
ozone (O_3), xx, 12–13, 22, 199, 201
  ozone depletion, xx, 12–13, 22
  ozone layer, 199, 201

P

pH, xxi, 49, 56, 58–62, 64, 67–68, 74, 77, 239, 261, 265
phosphorus, 57, 205
photosynthesis, xvii, 103, 107, 140
Pichia stipitis, 41, 52
plastic, xviii, xxvi, 18, 20, 37, 56, 73, 76, 90, 94, 98, 153, 161, 163, 175, 187–188, 215, 218, 241–243, 245–247, 249–253, 255, 257, 259–263
  plastic solid waste (PSW), xxvi, 241–243, 246, 249–252, 255–256, 259–260
pollution, xx, 3, 12–13, 15, 21–22, 25–26, 28, 36, 154, 186, 218, 249
  pollution abatement, 249
polyethylene (PE), 243, 246, 249, 255, 259–262
polyethylene terephthalate (PET), 243, 249, 255, 260–262
polymers, 107–108, 118, 242–243, 246, 256, 261–262
polysaccharides, 38, 107, 112, 141
population, xxii, 4, 7–8, 88, 100, 148–150, 153, 189
  population growth, xxii, 4, 153, 189
power plant, 21, 25–26, 191, 197, 199, 207, 215, 239
  co-generative plants, 157

Index

pressure, xxvi, 104, 126–127, 131, 134, 136–137, 139, 147, 154, 164, 174, 222, 227, 229, 232, 255–256
  partial pressure, 126–127, 131, 222, 256
pretreatment, xxi, xxiv, 36–37, 41, 44–45, 48, 50–53, 108, 148–149, 165–166, 168, 191, 208, 243
  mechanical pre-treatment, 148–149
proton, 112, 114
  proton transfer efficiency, 112
pyrolysis, xviii, xxvi, 4, 18, 101, 108, 110, 141–142, 190, 205, 207–208, 214–215, 220, 226, 239, 242–243, 251, 261–263

Q

Quantitative Structure Property Relationships (QSPR), 105
questionnaire, xx, xxiii, 11, 14, 16, 22–23, 26, 29–30, 89, 99, 265

R

reaction rate, 131, 221, 256, 259
recycling, xxiii–xxiv, xxvi, 12, 30, 88–89, 94, 98, 148, 172–173, 201, 218, 242–244, 246, 260–263
  thermo-chemical recycling, 242–243
refuse-derived fuel (RDF), 8, 26, 148
residence time, 110, 175, 249–250, 255, 259–260
resources recovery, 148

S

saccharification, xxi, 37, 39–42, 47, 49–52, 70

Simultaneous Saccharification and Fermentation (SSF), xxi, 40, 50–51, 56–57, 66
Saccharomyces coreanus, 41, 52
scrubbing, 218
sensitivity analysis, 202, 205
Separate Hydrolysis and Fermentation (SHF), 40
sewage, xviii, xxiv–xxv, 84–86, 189–195, 197–199, 201, 203, 205–209
  sewage sludge management, 190
  sewage sludge post-treatments, 190
  sewage treatment system, 189
shipping, 198
shredding, 218, 261
sieve, 76, 95, 98, 153, 161, 163–164
  sieve cut-off, 153, 161, 163–164
slag, 13, 15, 26, 158, 240, 251, 255
slippage, 218
sludge, xviii, xxiv–xxv, 19, 84–86, 165, 189–195, 197–199, 201–209, 252
  dewatered sludge, 190–191, 194–195, 197, 205–206
soil conditioner, xxiii, 16, 21, 27, 36, 87–89, 91, 93, 95, 97, 99
solid separation, 226, 232
solid waste (SW), xvii–xxvi, 4, 6–7, 11–12, 15, 24, 26, 28–31, 51–52, 68, 73, 84–86, 88, 100, 124, 126, 128, 130, 132, 134, 136, 138, 140, 142, 147, 149–151, 153, 155, 157, 159, 161–169, 171, 186, 189, 198, 206, 213–215, 217–221, 225–226, 228–229, 234, 236, 238, 242–243, 249, 260–262
solubility, 104, 117, 135–137, 139
sorting, 98, 218, 252
starch, 35, 39, 41, 44, 48, 55–56, 59, 104, 107
steam, xxv–xxvi, 52, 108, 112, 154, 156, 158, 164, 214, 218–222, 225,

227–228, 232, 234–236, 238–240, 242, 249–252, 256, 262–263
    steam injection, 214
    steam turbine, 154, 156, 164
    steam-fuel ratio, xxv–xxvi, 234–236, 238
stoichiometry, 229
sugarcane, 50, 215, 239
sugars, xx, 35, 37–40, 44, 47–48, 52, 85, 108, 110, 112, 114, 117
sulfuric acid (H_2SO_4), 48, 112
sustainability, 4, 147, 166–167, 207
syngas, xviii, xxv–xxvi, 3, 108, 214, 218–220, 233–238, 241, 243, 246, 251, 260

T

tar, 219, 238, 249–251, 255, 262
tax incentives, 3
temperature, xxi–xxii, xxvi, 4, 20, 49, 56, 58–61, 67–68, 74–78, 82, 84, 104, 110, 112, 114, 117–118, 122, 125, 131–132, 134–135, 139, 154–155, 158, 164, 173–177, 180–181, 184, 186–187, 191–192, 205, 207, 214, 219–221, 224, 227, 229, 232, 235, 237–238, 240, 243, 246, 249–252, 256, 259–260, 262–263
    gasifier temperature, 237–238
thermal
    thermal cracking, 243
    thermal cycle, 154
    thermal deactivation, 60, 62
    thermo-stability, 56, 60, 68
thermolysis, 242–243, 245
thermophilic, xxii, 75, 78, 83–84, 86
total organic carbon (TOC), xxii, 76–78
transportation, xxiii, 5, 91, 105, 107, 110, 140, 158, 196, 198

U

urban planning, 189
urbanization, 7, 87, 165

V

viscosity, xxi, 37, 39–40, 48, 104
vitrification, 4
volatile solids (VS), 38, 83, 133

W

waste
    waste characteristics, xx, 8, 10, 29, 88–89, 149
    waste composition, 7, 11, 16, 28, 149, 161
    waste feedstock, 3, 5
    waste generation, 88–89, 94, 100, 241
    waste hierarchy, 172
    waste management, xviii–xxi, xxiv, 4, 6–7, 12, 15, 29–30, 55, 88, 94, 99–100, 147, 149, 162, 166–168, 172, 186, 260–263
    waste material, xxiii–xxiv, 4–5, 88, 94, 104, 172, 174, 260
    waste production, 148–149, 158–159, 163, 165
    waste reduction, 88–89
    waste separation, 5
    waste treatment, 8, 10, 15, 30, 68, 118, 261
    wastewater treatment plant (WWTP), 19, 165, 207
water discharge, 156
water supply plants, 11
wheat straw, 36, 41, 50–52, 239

Y

yeast, 49–50, 52–53, 57
yield, xxi, 37, 40–41, 44–47, 59, 68, 110, 112, 130, 134, 156–158, 164, 166, 194–195, 228–229, 242, 250–252, 259

Z

zeolites, 108, 114, 118, 243
zero waste, 94, 100